being *me*

pete moore

being *me*

what it means to be human

WILEY

Published in 2003 by   John Wiley & Sons, Ltd, The Atrium, Southern Gate
Chichester, West Sussex, PO19 8SQ, England

Phone   (+44) 1243 779777

Email (for orders and customer service enquiries): cs-books@wiley.co.uk
Visit our Home Page on www.wiley.co.uk or www.wiley.com

Other Wiley Editorial Offices

John Wiley & Sons, Inc. 111 River Street, Hoboken, NJ 07030, USA

Jossey-Bass, 989 Market Street, San Francisco, CA 94103-1741, USA

Wiley-VCH Verlag GmbH, Pappellaee 3, D-69469 Weinheim, Germany

John Wiley & Sons Australia, Ltd, 33 Park Road, Milton, Queensland, 4064, Australia

John Wiley & Sons (Asia) Pte Ltd, 2 Clementi Loop #02-01, Jin Xing Distripark, Singapore
129809

John Wiley & Sons Canada Ltd, 22 Worcester Road, Etobicoke, Ontario, Canada, M9W 1L1

Wiley also publishes its books in a variety of electronic formats. Some content that appears
in print may not be available in electronic books.

British Library Cataloguing in Publication Data

A catalogue record for this book is available from the British Library

ISBN 0-470-85088-4

Typeset in 10.5/13.5 Photina by Mathematical Composition Setters Ltd, Salisbury,
Wiltshire.
Printed and bound in Great Britain by T.J. International Ltd, Padstow, Cornwall.
This book is printed on acid-free paper responsibly manufactured from sustainable forestry
in which at least two trees are planted for each one used for paper production.

Contents

Acknowledgements

The best way to look at what it is like to be a human being is to study individual people, to look at people's lives. This moves us away from theorising and into the real world. Doing this, however, means that you need to intrude into some of the more curious corners of an individual's history and personal stories, and talking to an author therefore calls for a mixture of courage and trust. The book owes its life to the people who have been prepared to expose their lives to gentle scrutiny, to the people who have revealed critical aspects of what makes them tick. Without the likes of Edward Bailey, Paul Bakibinga, David Barker, David Bird, Ann Jeremiah, A. Majid Katme, James Lovelock, Eileen Piddock, Jo Rose, Judy Tabbott, Christine Whipp, Arthur White and Rowan Williams the book would have been yet another polemic.

At the same time I'm grateful to those who gave me time and inspiration with various parts of the book. These include: Ann Broad, Paul Farrand, Mike Hawkins, Jane and Martin Hiley, Mike and Caroline King, Kenan Malik, Alexina McWhinney, Paul Sandham, Tom Shakespeare and Mary Warnock. The charity Changing Faces also very helpfully put me in touch with David Bird and Eileen Piddock.

As with all my books I have been supported, encouraged and goaded into action by Adèle, and the book owes much to the time given me by my commissioning editor Sally Smith – many thanks.

Introduction

Being yourself

There is nothing easier than being yourself, until, that is, you think about it. Few children running around a school playground would ever pause to ask what it is that makes them what they are. And on the whole, even fewer adults give it much thought, but every now and then through history there are explosions in interest.

Writing towards the end of the Renaissance, a time when the meaning of life was under constant debate, William Shakespeare placed the following lines in the depressed Hamlet's mouth.

> What a piece of work is a man! How noble in reason! how infinite in faculty! in form, in moving, how express and admirable! in action how like an angel! in apprehension how like a god! the beauty of the world! the paragon of animals! And yet, to me, what is this quintessence of dust?[1]

It is as if Hamlet starts with the list of things that he thinks define a human being and then runs out of faith in his own judgement. As many people before and after have discovered, the simple categories that we try to use to define day-to-day existence soon become dangerously thin when exposed to careful scrutiny. Using some physical description of who we are is problematic if a person is physically disabled, because it begs the question whether a human being could ever become a non-person. Using mental capability has similar perils – a brain-dead person could become defined as non-human, and therefore legitimately the subject of scientific experimentation.

As far as day-to-day living is concerned, the question of what it is to be me has no particular relevance. Quite obviously I am me, you are you and, who knows, one day we may meet. But probe only a little beneath the surface and you soon encounter troubling indicators that the apparent simplistic certainty may

---

[1] Shakespeare W. (1601), *Hamlet, Prince of Denmark*, Act II, Scene 2.

have a little to be desired. One of the classic pairs of questions that occupied much thought during the twentieth century was: When does my life start? and Can we define a moment when it ceases? Both of these demanded that we have a definition of what life is. Neither question was answered by arriving at universally agreed definitions, although politicians and lawyers did manage to create a few laws setting out boundaries in an otherwise poorly marked out and all too often earthquake- and volcano-strewn landscape.

I can remember as a mid-teen A-level student staring down a microscope in the school biology lab. In clear focus was the regular matrix of cells that makes up a section of a thin slice of onion skin. The specimen was stained so I could see various elements of the inner structure of each cell. These were complete cells that at one point had been alive, but now were dead. I stared at them for the best part of an hour and a half, trying to see if I could spot why these cells were dead. It seemed to me such an important question, but I felt that I would be laughed at if I raised my hand and asked it. So rather than submit myself to ridicule I just stared. The overall structure appeared to be there, and I squinted my eyes, racked the focus up and down and adjusted the light in an attempt to spot all the cell contents that my text book claimed should be there; few of the books being honest enough to point out clearly that many of the features could only be seen with an electron microscope – not a school "toy". What would a living slice be like? I pondered. Would I be able to spot the difference? The label on the microscope slide told me that this material had been dead and sandwiched between glass for a decade, so clearly all biochemical processes had ceased to occur, but there was no apparent loss of life to see. A bell sounded in the corridor marking the end of the lesson and I was no further forward – neither had I got around to drawing the blasted thing. Moore had once again failed to complete the task set before him.

In previous centuries, the solution to the issue of death seemed obvious. You watched someone. When they hadn't breathed for a few hours, their heart had stopped beating, and their body was becoming inflexible with rigor mortis, there was a strong indication that they were dead – in popular parlance, a stiff. Tests like holding a cold mirror above the person's mouth and nose to see if there was any trace of condensation caused by the shallowest of breathing could help spot the faintest traces of life. And in the eighteenth century doctors would slice into an artery to see if any blood came out – no blood, no life. Blood after all was believed to contain "vital spirit", the very essence of "you", so much so that early ideas of blood transfusion were frowned upon because they were assumed capable of transfusing one "person" into another's "body".[2] In short, you were your blood, and when blood stopped moving you were no more.

This assessment of human life was very straightforward, but it wasn't enough once medical technology had generated the possibility of transplanting organs from one person to another. For an organ, such as a kidney, to be donated it must still be alive when it leaves the original body, or the donated organ would be useless. Likewise a heart in a heart transplant, or a liver, lung, hand or – maybe soon – even a face. But it would clearly be unethical to strip people of these vital organs while they were alive. Taking a heart out would effectively mean committing murder.

So what happened was that life was redefined and new boundaries were drawn up. The argument was that these new limits were much more "scientific" in the way they were defined. Life, we decided, ended not when the body completely stopped functioning and started to decay, but instead when the brain ceased to function – when you were brain-dead. This is different from the point when the cells in the brain have physically died, but it is the point where their operation has become so weak or disorganised that there is no evidence of any effective function.

By defining the moment of death as the time when the brain's functional activity drops to a point that a physician can't detect it any more, doctors managed to create a window of opportunity.

[2] Moore P. (2002), *Blood and Justice*, Chichester: John Wiley & Sons.

They could now pronounce a person "dead" while the body was alive, and then remove undamaged, still living, organs. With remarkable ease, specialist medical care can support a person's body for many hours, if not weeks and months, after their brain has failed, enabling a recipient to be lined up, surgical teams congregated and the relevant organs transplanted from one body to another.

This definition of life grew from an unchallenged statement that when we lose brain activity we can no longer think, make decisions, interact with other people or display any form of consciousness. There is no sign of mental life. When doctors decide that a person has no residual unconsciousness and there is good reason to believe that there is no hope of regaining her mental ability, then they declare that she is no longer alive. She is dead. It is a form of reductionism, in which we are "reduced" to our brain function. Nothing else is important.

The fact that we make this statement while the vast majority of the 100 million million cells constituting a human body are still functioning normally points to the reliance we have on a simple equation; "I think, therefore I am" or, as René Descartes put it, *"Cogito, ergo sum"*. Indeed, in many cases most of the cells in the brain itself are probably still alive at the point of death; just not working properly.

The result has been a curious leapfrog effect in our understanding of what it is to be a living human being. The medical convenience of this definition has had a significant influence over the way we view ourselves and the ethical considerations that we find ourselves making. For most people it is an almost unquestionable dogma that without brain activity we are nothing. Using this as the foundation to their thinking, UK ethicists in the 1980s drew conclusions about the ethics of experimenting on pre-implantation embryos.[3] Drawing from the definition of brain death, their argument started with the

[3] Warnock Committee (1984), *Report of the Committee of Inquiry into Human Fertilisation and Embryology*, HMSO, July.

assumption that life equalled brain activity. It subsequently moved seamlessly on to say that if an embryo is too immature to have any nerve cells, then it is too early for there to be any consciousness. If an adult is believed to be no longer alive once consciousness has ceased, then the embryo can't have begun to be alive before consciousness has started.

The argument seems unassailable. But on its own it is dangerous, because it creates an atmosphere in which our understanding of human life is brain-centred, and soon we discover that the value and esteem that we give to individual human beings rapidly becomes based on some assessment of how well their brains function. This is then reinforced at a practical level, because people with agile minds are more highly valued than those without. A crude assessment of this can be seen in the way we pay wages, in that highly educated people, such as actuaries, lawyers and doctors, tend to be paid more than those who left school at 16.

You can see this brain-focused thinking in the writings of people such as American psychologist Steven Pinker. In *The Blank Slate* he discusses the problem of identifying when a human life starts, "The demand by both religious and secular ethicists that we identify the 'criteria for personhood' assumes that a dividing line in brain development can be found."[4] His statement is true if you have taken on board that assumption that "you" equals "your brain".

A quick extension of this thinking means that someone whose brain is severely damaged is, by that category alone, unworthy of the support and care that we would normally give other members of our species. Some have even suggested that babies born with no brains would make ideal candidates for experimentation, because in every other respect, they have perfectly formed human organs.[5]

---

[4] Pinker S. (2002), *The Blank Slate*, Penguin, p. 227.

[5] Shewmon D. A, Capron, A. M, Peacock, W. J, Schulman, B. L (1989), "The use of anencephalic infants as organ sources: A critique", *Journal of the American Medical Association*, 261, pp. 1773–1781.

In this case, having no brain equates to not being alive, and that leaves the individual with next to no value.

This, however, runs counter to the experience of many who live and work with people who have sub-optimal brain activity, either through injury or because they were born that way. They may not be "rocket scientists" but these people are very much alive, and often contribute greatly to their community, even though they may need lots of help.

Taking the brain as the sole key to human nature is where things start to go wrong. If this is the case, the only tool that we have to understand ourselves is our brain, and the root to finding out about the brain is science. This poses two problems. First is the conundrum that the brain may not be powerful enough to understand itself – but then again it may be. The second problem is more difficult to surmount. True science, if there is such a thing, should be purely objective. Opinion should play no part in directing an experiment or interpreting the data. In reality, opinion is highly involved. Consequently, any research that sets out to try to understand the nature of humanity will be strongly influenced by current attitudes within society. As Kenan Malik says in *Man, Beast and Zombie*

> The questions scientists ask about the world, and the interpretations they place on their data, are often shaped by cultural attitudes, needs and possibilities. In most scientific disciplines, the cultural context does not impress too deeply upon the scientific answers ... [But] when it comes to the science of Man, however, matters are different. Human beings are not simply objects that can be prodded and poked, measured and theorised about. We are also the subjects that do the prodding, poking, measuring and theorising. In other words, humans, uniquely, are both the subjects that create the science and objects of that science.[6]

[6] Malik K. (2001), *Man, Beast and Zombie*, London: Phoenix, p. 9.

This doesn't negate the idea that our brain is likely to be a critical aspect in understanding what it is to be human, but it does indicate the extent to which the issue is not going to be easy to resolve.

An alternative mantra gained ground through the latter part of the twentieth century: I am my genes. Famously Richard Dawkins' book, *The Selfish Gene*, claimed that our existence is an artefact of chance, an accident waiting to happen. He argues that the way materials work means that order can come into being spontaneously from chaotically moving atoms, and once that happened a material like deoxyribonucleic acid (DNA) was practically bound to occur. DNA is powerful because it can carry information and it can readily be copied, but it is physically fragile and, to survive, DNA needs to live inside a cell. Evolution then found ways of making ever more successful cells and ever more successful organisms, each, according to Dawkins, more capable of protecting its DNA. Ultimately you are not important; it is the survival of the genetic material buried inside each of your cells that counts. We now have a different form of reductionism – this time you are your genes and nothing else.

While elements of the mechanism of genetic selection are demonstrably true, a sad consequence of taking this line of thinking to its ultimate conclusion is that we devalue all living things to mere machines. Any appearance of purpose is a figment of the over-zealous activity of our accidentally evolved brains. It's not particularly attractive, and once again I believe that it runs counter to day-to-day experience. Even Dawkins is the first to admit that this assessment leads to the conclusion that genes will make their organisms (or survival machines as Dawkins refers to them) spend their times selfishly fighting their own corners. He does acknowledge that he is presenting things as he sees them not as he would like to see them; "I am not advocating a morality based on evolution," he says.[7]

---

[7] Dawkins R. (1976), *The Selfish Gene*, Oxford University Press, p. 2.

The problem with this is that there are plenty of people who have taken this idea one step further and created the discipline of evolutionary psychology. This argues that we are driven by our genes. It goes further, saying that our genes were shaped over millions of years of primeval cave-dwelling and foraging for food. The general ethos that underpinned Robert Winston's 2002 BBC TV series, *The Human Animal*, was that our instincts haven't altered since those days. According to this theory, we are cave-dwellers in jeans and T-shirts, or hunter-gathers in designer suits and expensive hairdos; our selfish genes are still in charge. The idea is that the evolution that gave us phenomenally capable brains and manual dexterity has left our basic drives and motivations untouched. It's an interesting theory, but I wouldn't like to bet my life on it.

A consequence of this gene-based reductionism has been the emergence of a new form of fatalism. I can no longer expect to affect who I am, because I am programmed by my genes. If my genes make me violent, then expect violence when I am around. If they make me musical or intelligent, then expect fruits of those attributes as well. At the extreme there is no point as a parent doing anything to try to influence or educate your child, because their genetic imprinting will be so strong that you are wasting your time, as well as money and effort. Psychologist Steven Pinker claims that 50 per cent of our behaviour is predetermined by our genes, and 50 per cent is instilled through personal experience, though he chooses to assign this totally to experience that occurs outside the family.[8]

As well as tying a noose around the neck of any aspiration for self-determination, this train of thought has led to another set of panics – the playing god panic. If I am my genes, then anyone who wants to look at them or make small alterations in them is messing with the "me" in the very core of my being. Gene therapies that seek to cure disease by inserting correct versions of faulty genes become a radical form of character manipulation.

[8] Pinker, *Blank Slate*, p. 381.

The conclusion is inescapable if we take the genetic definition of ourselves too seriously.

In his book, *Our Posthuman Future*, professor of political economy at Johns Hopkins University Francis Fukuyama takes another tack, by searching for the unique identifying feature that brings dignity to humanity. His solution is to introduce "Factor X". This he says is the critical aspect that raises humans from mere animals into beings worthy of superior respect.[9] It's a novel form of reductionism.

According to Fukuyama, X can be different things for different people. For Christians, Factor X = God's image, though getting theologians to agree on exactly what that means is difficult. The answer would be similar for Jews and Muslims who, having religions with foundations based on the histories of Abraham, share vital texts at the core of their religious bibliographies. For atheists, says Fukuyama, Factor X = the human capacity for moral choice. "Human beings [have] dignity because they alone [have] free will." Fukuyama points out that it is this notion of free will that caused the moral philosopher Immanuel Kant (1724–1804) to conclude that human beings should never be used as tools to satisfy another's needs – in Kant's terminology they should be "ends" and not "means". You can see Factor X either as an interesting notion, or an elaborate cop-out. Either it solves the issue or it muddies the water.

The limitation with Fukuyama's argument is that it is a fudge. His desire is to pinpoint the essential element that makes humans unique, because he cannot bring himself to accept that we are just another species of animal that has been formed as a set of randomly created machines with massive capability. He identifies the element that sets us apart from all other living things and calls it Factor X, but then says that Factor X is whatever you want it to be. It fits comfortably into a relativistic world-view, which suggests that there are no absolutes. But if there are no absolutes, an absolute Factor X probably doesn't

[9] Fukuyama F. (2002), *Our Posthuman Future*, Penguin, pp. 149–151.

exist either. It is an interesting idea, but it hardly achieves his goal of identifying the vital element that defines humanity and indicates why we are unique among life on Earth – some would even say unique among life in the universe.

The uncertainty of knowing what it is to be human has a knock-on effect. It leaves us unsure about our relationship to the rest of the living, breathing world of plants and animals. Quite clearly, human beings have superior intellect and power to all other organisms. While you can point to ant colonies as examples of complex order, or bee hives as communities where altruism plays a prominent role in behaviour, nothing comes close to the complex structure of a city, supplied with sewers and power, libraries and art galleries, factories and offices. Just look around and see the incredible effect our species has had on the landscape around you. Hardly anything you see will be genuinely "natural". Almost everything has been tampered with in ways that aim to make the environment serve the needs of the human race. The change is often so severe that many say that the environment has been vandalised by our species.

We look at the effect we have made and panic. If we are so much better than the rest of the living world, why are we making such a mess of it? Why don't we hand over its running to non-human species that we naively think live such peaceful, non-destructive existences? We long for the old days when we were kinder to our world and revere our cave-living forebears who we seem to think lived in closer accord with nature. In saying this, we make an intriguing mental shift. No longer do we see our ancestors as undeveloped savages, but instead they become noble cavemen.

In the twenty-first century the issue of humanity's place in the world takes on a new hue. At the end of 150 years of discussion about Charles Darwin's ideas of evolution, the notion that humans were in some way a novel species that was unconnected with the rest of the living world has crumbled, and the animal within has become increasingly apparent. But now the genetic revolution has taken this one step further. Genetics tells us that close relatives of almost all of our genes appear in animals as different as chimpanzees and mice. A large proportion even occurs in bananas and microscopic worms.

Rather than causing us to marvel at the interconnectedness of all living things, this has tended to fuel a period of biological flagellation, as we castigate ourselves for the arrogance that made us think we were any better than any other animal. We start to question our rationale for assuming authority.

Philosopher Peter Singer challenges the view that human beings have any superiority or unique value. He argues that all animals should be ranked according to their currently displayed intelligence, according to how well their brain is functioning. Those at the top of the heap he endows with the term "person" irrespective of whether they are human – i.e. members of the biological species *Homo sapiens* – or some other animal.

> The right to life is not a right of members of the species *Homo sapiens*; it is ... a right that properly belongs to persons. Not all members of the species *Homo sapiens* are persons, and not all persons are members of the species *Homo sapiens*.[10]

Moving on

I am convinced that it is high time we broke away from the negative and narrowly defined, reductionist views of what it is to be a member of the human species. It is time to remind ourselves of something that is so blindingly obvious we seem to have forgotten it. We are multifaceted beings. Our brains, our genes and the world we live within are all aspects of being human, and each individual's variation in these factors helps define what it is to "be me". The complex relationship between humans and other living things suggests that many of the differences that distinguish human beings from other animals may be best described in terms of features that vary in the degree of expression, rather than an

[10] Singer P. (1995), *Rethinking Life and Death*, Oxford University Press, pp. 202– 206.

absolute presence or absence. A few, however, may be exclusively held by our own species.

Place a prism in a beam of white light and see what happens to the light as it passes through. As the light exits on the other side it splits into a rainbow of colours. We like to think we can describe the colours and hence we give them names – red, orange, yellow, green, blue, indigo and violet. There is, however, no boundary between red and orange, the two merge. The difference in colour is due to the wavelength of the light with red light having a longer wavelength than orange. There is, however, no point in the spectrum where you can say that on the one side we have red light and on the other side, orange. But knowing which colour of light you are playing with is useful. For example, red light carries more energy than blue, so you can use red light to keep food warm and blue light in places where you want to see but introduce as little heat as possible.

We can treat the investigation of what a human being is in this way. It recognises that you can analyse different aspects of a person's make-up – different wavelengths within their spectra. Each time you try to assess the person you will look at different parts of the spectra one at a time. This enables you to see some of their characteristics and qualities, though it recognises that the person is always more than this single feature. In addition, the metaphor recognises that there is an arbitrary nature in the exact positioning of boundaries between categories. For example, there is no clear-cut boundary between a person's physical aspect and their sexuality. Equally the boundaries between spiritual, relational and social aspects of existence are at best fuzzy-edged. Giving names to each characteristic is therefore a way of trying to get our minds around the issue, rather than claiming to have created precisely defined concepts.

In my search to make sense of this, I have interviewed a number of people – real people. Some of them are names you may have heard of, others are living their lives without the glare of previous public exposure. I have worked to tell each person's story in a way that highlights one aspect of what it is to be human. As I researched the book, my conviction grew that you must never try to take any individual aspect on its own as a definition of what it is to be a human being. The chapter titles announce that we are historic beings, social beings etc., but this should not be taken to imply that the people acting as witnesses in each chapter are confined to these facets, just that I have used their experience and life story to illustrate one feature. Each person is a holistic amalgam of all the facets.

At the same time, the prism metaphor recognises that some of the zones marked out will be larger than others, just as some people find that specific areas of their being are more critical to them than others. The demarcations are also dynamic, allowing the relative proportion of these facets to change over time. At times of your life your spirituality may be more important than your awareness of your genetic heritage. A few years later this situation may be reversed.

What emerges is a picture of human beings that is varied, vibrant and is vastly greater than can be attributed to any form of simplistic reductionism. Adding all the stories together we will gain insights into the enormity of "being me".

CHAPTER ONE

an embodied being

I crawled around the M25 motorway that orbits London, sitting in one of the thousands of cars attempting to squeeze through the tunnel under the River Thames. Few cars contained more than one passenger, most of whom were staring blankly ahead wishing to be somewhere else, but at the same time glad that their heaters were working against the bitter December wind and fine rain outside. My destination was Unit 6, Woodford Trading Estate, Southend Road, Woodford Green – a northern suburb of London. I followed Arthur White's impeccable directions, all the time wondering quite what he would be like.

My reason for chasing him down was simple. From everything I had read and heard about him, Arthur seemed to be an extreme example of someone whose life has been shaped by aspects of his embodied being, his physique. Now no one would doubt that being human requires that you have a human body, but what I wanted to find out was the extent to which the specific detail of individual's bodies could influence their life – could influence their experience of "being me".

For some people the physical attributes of their body may fit the cultural expectations of what is seen as beautiful. In some African cultures it could mean that a woman gains the security of a wealthy marriage because she has successfully fed to the point of obesity and beyond. Her life and her character will be intrinsically affected by her physical size. In contrast, a woman in Western societies may gain fame and financial security by being thin in a way that stretches concepts of health. With these attributes she could rise to the peak of high fashion and stride down Western fashion catwalks, charging telephone-number-sized fees for a few hours work. Fame and fortune follow her physique and radically affect her self-esteem, though not always positively – there is many a catwalk model who tearfully admits to taking cocaine and other uplifting drugs to quash feelings of inadequacy and loneliness.

Glossy magazines shout from newsagents' heaving shelves, telling us that if we get the right image, if we take control of our physique, we can get the life we desire. We know deep down that this is partly hype, partly lie, but also that there is more than a grain of truth in it as well. Psychologists claim that you will form an opinion about someone within less than one second of your first meeting, and that they will be just as quick to judge you. Dressing right for the occasion and choosing colours that complement your skin and enable you to display yourself in the best possible light can influence decisions that mould your life. When searching through clothing shops you can choose vertical stripes to enhance the appearance of height, or horizontal bands to make you look less thin. The choice is yours. Image consultants have established an industry based on the power of self-presentation.

On the car seat next to me is a publicity shot of Arthur. He is clearly someone who knows how to tailor his appearance. The photo is in black and white, deeply shadowed so that the right side of his face disappears into the black background. His mouth is framed by a carefully manicured close-cropped beard, and his chin is outlined by the crisp white collar and immaculately worn black tie. His left eye stares straight out from under a gently raised eyebrow, in a way that offers an open challenge – "Want to take me on? I think not." Though in the case of Arthur, it wasn't his physical appearance that interested me as much as his physical capabilities.

You see Arthur is a power-lifter – not any power-lifter but the reigning world champion. Six months before I met him, he had won his sixth British title, and only a few weeks before my visit he had returned from Argentina with his third world title. At fifty-one, he was old enough to be the father of many of the other competitors. His heaviest official dead lift is 380 kilos; it's the British record and still stands from the time he set it in 1982. On top of that he's won the European title four times. He's a powerful man with a physical history – and that history includes physical violence.

I found him doing an afternoon shift looking after his wife's warehouse. Unit 6 was stacked with shoeboxes clad in Christmas wrapping paper and packed with simple toys – each one donated from some schoolchild and heading out to needy children around the world. I realised that one of them had come from my son. If Arthur's coat had been red rather than blue and his beard considerably longer, he would have been a dead ringer for Santa, keeping watch over the volunteer workforce as they packed cases to a background of Christmas carols.

Meeting him was almost an anticlimax. At 5 foot 10 and $17\frac{1}{2}$ stone, he wasn't as physically intimidating as I had anticipated – but then I suppose I am 6 foot 2. And when we shook hands my hand came back unscathed, the bones uncrushed. But there was no doubt in my mind that I had just put my hand into an open vice. It had been his choice not to close it.

We moved to the quiet of an upstairs office and Arthur unpacked some of his past. His body had always been a key asset in life, and as a child he soon realised that it was built to win. Aged 14 he was chosen to be part of his school's athletic team. It wasn't a prestigious establishment, but Fairmead Secondary School soon had a runner to be proud of. Arthur's speciality was sprinting and he enjoyed sports more than anything – sprinting is all about power and gritty determination. By the time he was 16 he had raced to become the West Essex Junior Champion at 100m, and then Essex Junior Champion, equalling the national 100m record.

Lifting weights is part and parcel of any training programme in athletics and is particularly important in sprinting where pure explosive power is critical to success. "I started weight training purely to assist the athletics that I was involved in – just to get some strength. Things drifted on from there really," commented Arthur, but the drift was impressive.

Arthur's body wasn't particularly big; in fact it has never gone over 18 stone, which for a heavyweight lifter is modest. But he found that it was strong – the genetic heritage underlying the way his muscles operated had provided him with a powerful legacy. "As my body developed I felt good about myself. Call it vanity, but it gave me courage when I started looking at the opposite sex, though the first time I saw the girl who became my wife I became shy," he chuckled.

In 1968, aged 17 he left school and set out to conquer the world. Within three years he had a job as a carpenter and had married Jacqui. As with many school sports victors, the reality of work meant that time was pressed and weight-training was relegated to the status of a hobby, something to do as a wind-down at the end of a busy day. All the same, Arthur started pumping iron at one of the more serious gyms in Forest Gate, north London. Run by Wag Bennett, it was frequently used by major hard-core body-builders including the likes of Arnold Schwarzenegger. "I met a weight-lifter there who was Commonwealth Champion and was in the '64 Olympics, George Manners, British Champion, and he took me along to the Bethnal Green Weight-lifting Club, which is where I trained for abou' 18 or 19 years," Arthur explained.

Training became less of a hobby and more of an obsession, lifting weights for a couple of hours five days a week. The training was carefully organised so that smaller muscles got the attention they needed and techniques were practised thoroughly. To compete in power-lifting, athletes need to perform three exercises: squat, bench and dead lift. It's the accumulation of the three that gets you the total to win overall. "In competition it may be that you weren't the best squatter or the best bench or even the best dead lifter, but the three added together would give you the gold medal."

"The glory of a young man is in his strength, says the Bible," smiled Arthur. He knew this more than most. While many youngsters pretend they would like to be the next Arnold Schwarzenegger, Arthur was on his way. "You get a role model, not necessarily the right one, and it does build your self-esteem, you know. But then again any sport I think, or anything you are good at, if you develop it, does build you up," he said.

Arthur loved his body. He loved checking how it was developing in the mirror-lined gym walls, and it was shaping up nicely. The muscles were clearly defined as they rippled beneath his skin. It was just as well that he had been born with exceptionally strong tendons and bones, otherwise he might have torn himself apart, or crushed himself as he worked out.

His body also affected his mind. He began to have vivid dreams of using his unusual strength in heroic situations and started to fantasise about being invincible. He built up scenes of train crashes in which he alone survived the initial impact unscathed, and used the power of his body to haul railway carriages off fellow travellers. On another occasion his fantasy involved him climbing from a wrecked car only to find an enormous piece of metal spiked though his body. To the amazement of onlookers he simply pulled the offending item out, threw it to one side and strolled off. Nothing could touch him.

By 1975 he was competing regularly, breaking records in British, European and Commonwealth competitions. He was also self-employed, running a large building contractor's and had a baby daughter. Arthur's physique had now hugely affected not only himself, but also those around him. Life was hectic, but life was good.

Six years later Arthur was lifting for Britain all over the world, travelling throughout Europe, North America and India. He made opponents tremble and gained celebrity status. He won the British Championships, the European Championships and came second in the World Championships.

Work and family commitments kept Arthur out of competition for a few years, but in 1985 he again challenged for the European title. This time he came third, two places behind his ambition. Being beaten was not part of his plan and Arthur was frustrated that his body had let him down. He became more anxious about his body when a growth appeared in his throat. It needed to be surgically removed. Then his father died. Reality cut in. These things called bodies weren't as infallible as he had started to believe. He became vividly aware of his own mortality.

Arthur's desire to win became a desire to win at all costs and soon he had found a source of assistance – chemical assistance – steroids.

I think the people you mix with have an influence on your life. I was training in many different gyms where drugs were rife. They were easily obtainable, relatively cheap, and you think it's a quick and easy way of getting to the top of the sport you are in. So I just went along with the crowd. Everyone else was doing it, so I was doing it. I was fuelling my body with steroids, I was running my own business, training hard, obviously I had a family to look after and I was getting mentally tired.

Combined with the packed lifestyle, the steroids started to cause mood swings. Arthur combated this by taking amphetamines, and then added cocaine to the cocktail to boost his energy levels. Now he could work hard enough to run his company and pay for his house in England and a villa in Spain, and do his training.

But working as hard as he could still left him with not enough cash to fuel his drug habit. At £200 a week, paying for pills, powders and injections was beginning to hurt – and it all had to be found without his wife knowing, and she ran the company books. A weight-lifting mate suggested a solution – door work. Being a bouncer could get you anything up to £400 a night if you were big enough, looked mean enough and had the right contacts. Arthur's body had taken him a long way from the youthfully ambitious schoolboy sprinter who simply thrilled to win.

It was remarkable thinking back to how doing drugs changes you. Steroids change your psychological outlook on life. Basically they are hormones. Pumping yourself full of strong male hormones causes an imbalance – it has got to be wrong. The psychological outlook is phenomenal, it just goes AWOL, it goes absolutely AWOL, and it had a massive effect on me, and my life, you know. And everything around it, my wife, my family, my kids.

Now the consequence of Arthur's physique took him into new territory. The world of fanatics' gyms is no place for the faint-hearted, but security work at notorious nightclubs is guaranteed to get your adrenaline running. Being nervous at first, he soon found one unexpected advantage. Not only was he getting paid, but also as part of the job he frisked punters at the door. This gave him a valuable cache of drugs each evening. The work boosted his income; the confiscated drugs reduced his outgoings.

On top of this, he also started to love the violence that came with the task of evicting troublemakers. He knew that he was stronger than any of the punters and that he could win any fight; the heat of the moment adding bravado to his strength.

It had now been a few years since he had competed, but Arthur decided it was a good time to grab another title. He walked cockily into the 1987 British Championships and came second to someone he felt he should have beaten easily. This spurred him on to more serious training and later in the year he won silver in the World Championships.

But long-term use of drugs wrecks your body and ultimately destroys your life.

> They nearly killed me. I left my wife; had an adulterous affair; ruined my business; I lost everything – homes, cars, money, family – the lot. The drug addiction was killing me. I had an X-ray done in 1992 and it clearly showed that my heart was ballooning to the size of a football. Steroids harden arteries and cocaine tries to pump the blood around too quick – you don't really need to be a rocket scientist to know that something is going to go wrong. Nine guys that I have known, all under the age of 40 taking steroids and cocaine, have had massive heart attacks. One guy in particular was called John Paul Sigmason, an Icelandic fellow who was three times the World's Strongest Man. It was all over the television. He was 32, 6 foot 4 and 23 stone, and dropped dead while training. Steroids and cocaine ... Heart exploded ... And that's the way I was going.

At one point Arthur realised that there was another way his body could earn money – serious money. He started running an illegal debt-collecting business for the big men in east

London. It was a business that used violence as a basic currency of persuasion.

> I was training harder then – I was probably in the best physical shape I had been in in my life. I was fit and I was strong. I wasn't healthy because obviously the drugs were causing me damage, but I was fit and strong, and my physical appearance was very important. When I stood there they knew that they was going to have a handful in front of them. And yes, it was very important. I remember stabbing a guy twice and attempting to cut his ear off. Yeah, the lifestyle was one of violence fuelled by drugs.

Unsurprisingly a drug test in 1988 caught him and he was banned from competition for a year. "I then went to South Africa to compete and for that they banned me for life." His ego was such that he felt he was above the Gleneagles Convention, which banned all sporting contacts with South Africa as a way of putting pressure on that nation to abandon the racial discrimination inherent in apartheid.

Arthur was heading for rock bottom. A couple of failed suicide attempts didn't help the situation either. "I was involved in nothing but drug-powered violence and that is when I asked God to come into my life," he explained in a matter-of-fact manner.

> I chose to ask God to come into my life. There was no booming voices or opening of the heavens or choirs of angels and I really did want that to happen. I wanted the sign, the finger to come out of heaven and point me out. It didn't, but I knew something happened. The paranoia went; the fear went; I threw my knife away that night. I stopped my drugs that night. It wasn't easy over the following months, but there was something there that helped me through. I can't deny it. I don't exactly understand it all but I don't understand how your microphone works and your little minidisk player, but we all accept it and that's the way I feel about my faith and my spirituality.

It is his explanation for the radical change that occurred over the next few years, culminating in him getting free of drugs and reuniting his family.

The end? No. This born-again muscle man still uses his physique to do some persuading. Now rather than trying to get frightened people to part with their cash, he works for Tough Talk, a registered charity consisting of a group of men who each claim to have had a "Damascus Road" religious experience after a rough start to life – to have seen the light. "We give weight-lifting demonstrations, tell our life stories and bring a gospel message. It's this work that spurs me on at the moment," he explained. Arthur has learnt through life that standing there with a powerful body affects people. It gives him confidence and it influences their thinking.

But this wasn't enough. "I wanted to win a world title drug-free," Arthur said. The reason was twofold. First was the sense of satisfaction that would come from winning without cheating, but more importantly Arthur felt that he would have no ability to stand before people if he was only an ex-lifter who injected his way to the top. His credibility demanded a clean win – it needed him to prove that his body was one of the best. Then he would be a real man, a man to be listened to, a man to be taken seriously.

So, three or four years ago he approached the British Weight-lifting Association and told them what had gone on in his life and how it had changed. He explained that he needed some credibility, some physical status. They reinstated him and Arthur was soon back representing Great Britain. "I would like to think I am a good ambassador for the sport as well as my own faith." Arthur claims to have been drug-free now for 10 years and enjoys saying, "Look, I am World Champion, I am competing in the same organisation, against the same men I was 20 or 30 years ago, and I am now drug-free and I am winning." He adds, "I would be lying if I said I didn't enjoy the competition; I do enjoy the competition, it's great. I don't enjoy the training so much now, but again that's age."

"Lots of people train who are never going to win – why do they bother?" I asked.

"Vanity, vanity, vanity. Without a shadow of a doubt. Vanity and ego," he replied.

When you go into an average gym and you see the guys training there, some of them have got very nice physiques but would never compete, ever. They would never make Mr Olympia or Mr Universe, but will continue training, continue body-building, continue taking drugs. It can only be vanity. It must stem from the sexual thing. You know it's the male/female thing. We want to make ourselves more attractive to potential partners.

Again, I think a lot of it is linked to age. You know, vanity was part of my life. The shape I was in, the condition I was in, what I looked like. I think as a younger man I was more concerned with fashion and what I looked like, but as I've got older it's not so important. Certainly I had a huge ego, a huge ego.

We all like to think we look presentable and acceptable. I wouldn't say I am the most beautiful man on two legs, far from it, but I like to think my wife doesn't have to put her hand over her eyes laughing her head off, and vice versa. We see things in the physical; that's the way that we are made. It's how someone comes across to you; first impressions, no matter who it is, whether it is going to end in a relationship or whether it's a passing relationship, it's how you see someone.

Facing yourself

If you are ever in doubt that we live in a world where appearances count, just take a moment and scan the covers of glossy magazines in a newsagent's. Most will display a stunningly good-looking person in seemingly perfect health. Teeth lined up in perfect harmony with each other and glinting in the light of the flashbulbs – or painted on afterwards by computer. The editors put them there because they know that being reassured by the image, we will be encouraged to buy the publication. It's a technique that is not lost on the money-makers in big business. Flick through any financial end-of-year report and there again are the same reassuring images – faces of the directors and managers, sitting at pristine desks in clinically clean offices. Again the photos are immaculately lit so that the end result looks more a photo of a supermodel than of a finance guru. The idea is that the face looks so strong that you trust them to protect your money, so honest that you would never doubt their motives, and so friendly that you wish you'd met them years ago and had been given the opportunity of "coupling".

It all relies on the well-known fact that we judge people by their faces. In Western societies a face that is equally divided in thirds is counted as beautiful. The first section runs from the top of the head to the eyebrow, the second from the eyebrow to the tip of the nose, and the third from the tip of the nose to the bottom of the jaw. There should also be impeccable left-right symmetry. "This pattern of beauty has been around for a long time, just look at the statues of Greek goddesses – perfect symmetry in the face," Dr Paul Farrand explained when I met him in his office at the University of Plymouth in Exeter, England. Farrand is a psychologist who has studied the effect on people of having unusual faces.

Taken on its own, you could be forgiven for asking, So what? But the issue is that for well over a century people have been dabbling with the idea that the face is more than a mask, that it reveals critical information about a person's character. One of the most persistent researchers in this area was Francis Galton (1822–1911). This cousin of Charles Darwin became a pioneer of eugenics. He took hundreds of photos of people convicted of various crimes and analysed them, determined to produce a method of classifying them in terms of criminality. His "criminal type" of person tended to be shorter than average, had a broader neck, an enlarged forehead and stunted face.

While the detail within Galton's work is now looked at with more than a measure of amusement, the underlying concept has not gone away. We make judgements about the sort of behaviour we can expect from others by looking at their faces – and intriguingly that can influence the person they are, or the person they become. There is the moment in the Marx brothers' film *Duck Soup* where Mrs Teasdale says, "He's had a change of heart." To which Groucho in the guise of Firefly replies, "A lot of good that'll do him. He's still got the same face."

"There are some studies in the 1970s and 1980s of classroom teachers," Farrand told me.

To start with researchers rated the faces of the children in terms of beauty, then they looked at the teacher's behaviour towards the children. Those who were perceived as uglier were punished more frequently and punished more harshly than those who were rated as more attractive.[1] So two children with much the same type of behaviour were treated differently on the basis of their face. You can see how one child might start to rebel, and then start living up to their label. Label someone as naughty and they start adopting the characteristics of naughty children.[2]

At the same time, less attractive children are more likely to be the victims of bullying than age-matched attractive peers.[3]

Some states in 1950s-America took this one step further and adopted a policy of performing plastic surgery on criminals.[4] The idea was that people became criminals because they were responding to society's stigmatisation of them as a result of some

---

[1] Rich J. (1975), "Effects of children's physical attractiveness on teacher evaluations", *Journal of Educational Psychology*, **67**, pp. 599–609.

[2] http://www.getting-on.co.uk/toolkit/mp_intro.htm

[3] Lowenstein E. (1974), "The bullied and non-bullied child", *Bulletin of the British Psychological Society*, **31**, pp. 316–318.

[4] Lewinson E. (1974), "Twenty years of prison surgery: An evaluation", *Canadian Journal of Otolaryngology*, **3**, pp. 42–50.

feature of their face. Receiving a normalised face gave them a new start in life. The ethics of the method are at best highly dubious, but the researchers claimed that it reduced the re-offence rate. In itself it gives lie to the adage "give a dog a bad name". This surgery suggests that people can change if society gives them a fair chance, and in this case giving a fair chance requires that we don't treat first impressions as if they are the full story.

Looks also have a tangible effect in the job market. "We investigated a sample of over 11,000 people aged 33, examining the effects of looks, height and obesity on hourly pay, employment and on a person's marriage prospects," explains Barry Harper from the London Guildhall University in a paper summing up some of his research. "Looks were assessed when these individuals were still at school." His results showed that the penalty for unattractiveness in men is a 15 per cent reduction in their pay package, and a 11 per cent drop for women.[5] He also found that tall women earned 15 per cent more than did shorter ones, but tall women and short men were less likely to get married, although being overweight and male had no influence on marriage prospects. He concluded that "although there is some variation between jobs, the effects of appearance are generally widespread suggesting that they arise from prejudice and in particular, from employer discrimination".

Born different

If first impressions are so important, what would be the effect of being born with an unusual face? I had been given a phone number for David Bird, a man with a "vascular anomaly" surrounding one of his eyes. I called to arrange a meeting. He was evidently a friendly, outgoing, bubbly sort of chap and my first impression of him was that he would be the life and soul of any party he went to. As he signed off I had the first inklings that

[5] Harper B. (2000), "Beauty, statute and the labour market: A British cohort study", *Oxford Bulletin of Economics and Statistics*, 62, pp. 773–802.

that might not be the case. My plan was to take an early morning train from London King's Cross to King's Lynn, the nearest station to his home. He would meet me there.

"Right, then, I'll see you tomorrow," said I.

"Great," he replied, and then added, "but, how will I know you?"

I hesitated, and started mumbling something to the effect that I was 6 foot 2 and would probably be wearing a large black overcoat ...

"Oh, well," he interrupted, "you'll recognise me. I mean, I stand out a bit."

We left it at that, but as the train pulled into King's Lynn station I started to wonder quite what he meant. I stepped out onto a fairly empty platform and started looking around. Immediately opposite was a man with a large-brimmed hat, pulled down so that his face was shaded from the bright winter sunshine. I moved forwards and he looked up. His face was heavily wrinkled, with a minor scar running across his cheek and drooping bags under each eye. Well, there's a strange face, I thought, but then paused. This was the face of everyone's grandfather, well-worn after years of service and exposure to the elements, but nothing particularly remarkable.

I turned and moved along the platform, noting that to an extent I could have been here to interview any one of three or four of my fellow travellers. But David was right. I did recognise him when I reached the ticket hall. While everyone else's faces had more or less signs of wear and tear, David's was distinctly different. Where his right eye should have been was a prominent mass of skin and bulging purple veins. As I came closer I realised that his right eye was there, but was almost an inch lower than his left and peered through almost closed eyelids.

He strode confidently towards me, his hand extended, his head up and his face sporting a disarming, if fractionally lopsided, smile. "Nice to meet you," he said.

Walking through the town centre was a mildly curious experience. We'd turn a corner and meet someone who would do a double-take as they saw David's face. Some would instantly look away and pretend that the encounter hadn't occurred, others would stare with a mixture of disgust or distress stretched over their faces. One woman took an involuntary step back. David, however, walked tall, seemingly perfectly at home with himself and the reaction that his presence caused to others. We took a table in a town-centre coffee shop, moving to a quiet corner not to get away from people's stares, but so that I was in a better position to record the conversation.

"I was born with this," said David, "though it was never as bad as this." He pulled a childhood photo from an envelope. The child who stared out was cuddling a traditional teddy bear and grinning self-consciously for the camera. His right eye socket was slightly too large, but relative to the situation now, it didn't appear to be desperately noticeable. Even so, by the time he was three months old his parents were already taking him to doctors and hospital specialists. "Then I got hit in the face with a rounders bat one day at school when I was seven. We were in the schoolyard. The bell went and this boy flung the bat over his shoulder and I was behind him. It were just fate," he explained in his gentle Norfolk accent, without any sense of anger. "It done so much damage that it made it a hundred times worse. I remember sitting in the classroom sobbing, worried that my mother was going to go crazy at me over what had happened."

At home his mother did go crazy and rushed David to his doctor, but there was nothing much they could do. From then on there was no hiding the fact that David's appearance was seriously different from most other people's. Forty years later the orbit behind his eye is considerably larger than it should be, causing the eye to rest lower than it should do. The socket is also packed with blood vessels with the result that his right eye and the area around it is pushed forward. "It's never going to get any better – it just gets worse," he added. "It's unusual, but I've got it for the rest of my life, so I've just got to get on with it."

David had been born on Watering Farm, a remote farmstead in the Brecklands of Norfolk. A few months after his arrival the family decided to move three miles into the local village of Thompson, and it was here that he grew up over his early years. His story is intriguing because it shows the ease with which people get used to the differences in each other's appearance, but the extreme reactions that we can all have towards people who at first sight appear to be different.

"As a little child growing up I didn't even know I'd got it – I wasn't aware of it," he told me.

I lived in a village with people I knew. I was brought up with all the village kids who just accepted me for who I was. I was normal like everybody else, you know. My family never made an issue of it – my friends never even say anything about it. Even now, people who know me just talk as if it wasn't there. It's not there. I'm me. I'm a person – they talk to me as a person.

While he was still young, David's parents managed to gather enough money together to take over a tenancy of the Wagon and Horses, a public house in the local village of Griston. This was a superb place to practise meeting people in that there were always workers and customers coming and going, but most were local and knew David so they made no issue over his abnormal appearance. "I had to meet people – I suppose that was good for me."

But then, when he was about 16 and had just left school, his parents moved into the Princess Royal, a pub right in the centre of King's Lynn. They hoped that living in the town would make it easier for David to find work. "That was a crisis. Everything coming at me all at once," he said, looking down.

That was horrible. Nobody took any notice of me in the village. Even in the local small town they all knew me. I'd never come to a big town before, but it were dreadful. I used to walk down the street with my hand over my face. They were dreadful years. One of the worst things I remember was going to discos with my mates and they would be up there with all the birds and there would be me standing watching – nobody wanted to dance with me – ugly little git standing over there in the corner. All your mates have girlfriends. No one wanted to know me. But that was part of my learning.

In the larger setting of the town it was possible for people to recognise him without ever getting to know him. This made him an easy target for bullying taunts and gibes. Schoolchildren tended to be the worst, but were not the only verbal assailants. Being seven years older than his brother, David frequently stood at the school gate waiting to bring him home.

The kids would come around and call me names, laugh at me and take the Mick out of me. Teenagers would have their say as I walked down the street, and lots of people would look at me as if I was something out of a horror movie. I had never been treated like this before and it knocked my personality. It affected me. It made me very introverted. I found it hard to look at people straight in the eye for fear of their rejection. It really knocked me. I were quite an outward-going little chap when I lived out in the country. Come to a place like this – that was quite a nasty shock. I just tried to keep in as much as I could. I was frightened to go out.

For a decade David tried everything he could to change his appearance in the hopes of reducing the ridicule and increasing his appeal to women. This included repeated rounds of surgery.

I needed to have it done. I wanted to be a better-looking person. Everything they flung at me, I'd have it, hoping that it would do something for me – hoping it would increase my chances. I had 14 operations. I went on a few one-night stands – I took a couple of girls out. They were all right, but I knew that in their minds they didn't really like me for what I had got – they didn't want me for what I was.

Self-esteem also comes from a sense of achievement, and David was aware that he was not achieving; instead he was being supported by Social Security payouts. "I used to feel embarrassed going to the post office to draw out that money. I wanted to make my own way," David explained. So not wanting to live on government handouts, he has worked in a succession of jobs, not all of which have been kind to his eye – lifting heavy weights increases blood pressure and can make it worse. Many of the jobs, however, put him right in front of people and gave him the opportunity to regain some self-confidence. Managing the wines and spirits section of his local branch of Tesco's supermarket was a challenge, but one he learnt to handle.

On the August bank holiday of 1983 he met Joanne. He had been invited to a barbeque party and almost didn't go because his elderly little dog was very ill. After persuasion he went. "I was introduced to Joanne and things went from there on. She had been born profoundly deaf and also knew what it was like to be lonely. Joanne took me for the person I was and did not judge me by my looks." On 25 July 1987 they got married and 300 friends and relatives packed the public hall to join in the celebration.

Even as he got married, David was all too aware that not everyone had the ability to see through appearances.

Getting married was such a wonderful feeling – I wanted to tell the world. I worked at a factory and I would pass this woman every day. I didn't know her, but we were passing strangers, so I thought I would tell her about the wedding. She turned around and said, "Oh, I didn't realise that anyone would want to marry a person with something like that on their face." That really took me back a bit. I thought a few things about her, but I didn't say it. She was severely overweight and wouldn't have made the front cover of a magazine, but she was prepared to pass comment.

Previous to meeting my wife I wanted to have everything done to make my face look better. But as I get older and have settled down with my wife, things don't worry like that any more. It's not important.

In fact, it is so "not important" that David could now only be described as an extrovert. "I'm terrible," he laughs, "you can't stop me once I start. I walk down the street, we come into town, there are people I know, not by name, through my parents being in the pub business, or my work in Tesco's ... Someone says, 'Hello, how are you?' My wife complains that we are there to go shopping, but I love to talk."

Clearly David's personality has been shaped partly as a result of his physical appearance, and by the way that people have related to him. You could argue that his underlying disposition is to be an extrovert, but his face has at times stood in his way and formed a barrier between his bubbly interior and the people he encounters.

"My face is me. It's what I've got, and if someone can't accept me for what I've got then that is their problem," says David with a sense of defiance.

But he goes further than mere reluctant acceptance. "I am happy with my face," he says.

A few months ago someone asked me whether, given the chance, I would change my face and I said, "No". The way I look has been more of an advantage to me. That's the way I feel. I have made an advantage of it. I think I have gained more friends with what I have had ... I'm just me, people have to take me as I am.

This is not naivety. Far from it. David is well aware that our current culture is one that places a lot of power on appearance. There is, after all, the tendency to see the Hollywood film goddesses with immaculate faces as normal, and for people to think that any deviation from these immaculate icons is a slur on your character. It is a fashion sensation that has given rise to the high-street availability of injections of botulism toxin, which paralyses facial muscles, relieving tension and effectively ironing out wrinkles. "For perfectly normal-looking people – I just don't understand it," says David. "I don't think they realise what they are doing."

Over Christmas 2002 he appeared on huge posters as part of a UK-wide advertising campaign run by the charity Changing Faces. Some of these posters appeared in Oxford Street right next to huge images of feature-perfect supermodels. On a TV chatshow,[6] Changing Faces director James Partridge explained that the campaign wasn't trying to deny that image is important, but instead to expand people's minds a little so that they realise it is OK to look different. James himself is well aware of the issue having had 55 operations aimed at rebuilding his face after a car crash when he was 18.

"I'm not disputing that people want to look their best, but these posters showed the other side," added David.

> I wanted to do the posters because I want to make people more aware that normal people have different-looking faces – it makes people more in tune with things. I'm a human. A lot of people prejudge you when they look at you. But I'm a person. I have a name. I have feelings. I want to be friendly. I'm not different. Like them, I'm human.

According to Partridge, the key problem that people who look different have is coping with the first few moments of contact. It is as we encounter someone that we have never seen before that we make our initial judgements. Working with Changing Faces, he is trying to help people find ways of engaging each other in conversation so that they move rapidly through those first few minutes. Once that is complete, most people find that they are no longer concentrating on the unusual facial feature, but instead communicating with the person. "They then see the real you," says David. Certainly, I found that the longer I spent time with David and talked with him, the less his face seemed unusual. Even after only an hour in the café and a couple of mugs of

[6] *Open Doors* with Gloria Hunniford, broadcast on ITV in December 2002.

coffee, I was starting to forget why I was there – already the warped side of his face was becoming less relevant.

Another example of this can be seen in David's relationship with his daughter. Lindsey was born in 1991 and had never known her father to look any different. For her, his unbalanced face is normal. This also applies to her school-friends who have grown up alongside David's family. "I've asked Lindsey how people at school react to me, and she says that they make no comments. But then they have grown up in the village where I now live, and they know me."

David sees himself very much as a wholly normal person behind a slightly unusual face. As such, he would probably argue that his appearance is not part of his personality – he would say that his face is almost a mask that prevents the casual observer spotting the calmly outgoing and humorous person behind the veil. While this is true to an extent, it is not the whole story because, unlike a mask, David can't take his face off. The consequence of living with it 24 hours a day, 365 days a year is that he has had to adapt his character to take it into account. It seems that people who are different in some way have two options. They can either withdraw and hide or they can practise techniques that allow them to become more assertive and stand tall, and at different times in his life, David has tried both.

For David, the assertive, open character has won over. "I'm such a strong-willed person," he says. "I've had to harden myself – in early years I used to get down with it, I felt quite lost and questioned who I was. It used to really hurt me. It played on my mind. You have to shut the problem away. When I walk down the street I'm a normal person. I don't have a problem."

Through his life, David has learnt a lot of strategies that have enabled him to gain control of situations that were initially threatening. It is a learning process that has liberated him, but has at the same time undoubtedly affected who he is. With appropriate help, the unusual nature of his face has become a tool that has shaped his personality. The process has taken time, and as a consequence David is grateful that he has had a lifetime to take it all in. "It must be dreadful for people who have nothing wrong and then suddenly – bang – their face is changed for ever."

A tumour too far

But that was exactly what happened to Eileen Piddock, who was born in Folkestone in May 1938, and has spent most of her life living in Kent. "The day started just like any other," she explained when I met her and her husband Alan in their bungalow nestling in a village just behind Dover castle. "It was 1971 and my husband had gone off to work. I rushed around getting Andrew who was ten and Sharon, aged eight, off to school. There were the usual pleas: 'Mum, I need money for the school trip.' 'Where is my PE kit?' 'I can't find my homework.' 'My shoes hurt!'"

With the kids out of the house, Eileen ran to catch her bus. She worked as a domestic in a home for retired naval seamen. Having just managed to catch the bus, she thought her problems were over. But getting off at the other end Eileen felt slightly drunk. "I couldn't walk properly. It was just as if I had had too much to drink," she explained. At first she wasn't alarmed and laughed it off. But her balance still hadn't recovered a few days later. Then she started being sick in the mornings. At this point she panicked – was she pregnant? But in an age before pregnancy tests, this was difficult to determine and there were no other signs of pregnancy. Instead, she was bumping into things and getting tired very easily. She started taking headache pills to tackle the aching pains in the back of her neck.

Even so, her doctor saw no reason for alarm. He kept saying that it was all in her mind, and that there was nothing wrong. "Women of your age, blah, blah, blah," laughed Eileen, though apparently she didn't laugh at the time. Alan took the doctor's line, and started telling her to snap out of it. Eventually her doctor relented and sent Eileen to see an ear, nose and throat specialist, but even then the tests showed nothing. By now Eileen was often having to crawl around on hands and knees.

In May 1971 the family bought a Mini car and went on holiday. "It was the first real holiday we had been on since having children. We went to Butlins at Minehead," she recounted. Straight after the holiday Eileen went to the hospital in Chatham for more tests before being transferred to St Bartholomew's Hospital in Rochester.

> There, by chance, a doctor from the Neurosurgery Department at the Brook Hospital was visiting another patient in the ward. He happened to see me walking along a corridor – I was all over the place. He asked to examine me. "You are very ill, my dear," he told me in a matter-of-fact way. These are not normally the words you want to hear, but to me it was sheer relief. At last someone believed that I was not putting it on.

Eileen had a tumour on her brain stem, and was soon on a ward in the Brook Hospital in London, a centre of excellence for tackling brain damage. Diagnosing the problem only raised another. To operate left the real possibility that Eileen could be paralysed all over. Not to operate would mean that she would most probably die within three months. Having decided on the operation, Eileen went home for the weekend. "I remember going to the Sunday morning service at the church where Alan and I were married, silently pleading with God, not to let me die."

In a short series of operations the surgical team managed to remove most of the golf-ball-sized tumour, relieving the pressure on the brain. This in itself should have made her feel better and at first all seemed well. The surgeon was overjoyed that Eileen could move her arms and legs, but the day after the most adventurous operation there were problems. Removing the tumour had damaged the nerves, and some of those damaged fibres drove muscles in the left of her face.

Fearing that Eileen would be terrified of her own appearance, the doctors and nurses made sure that she couldn't see a mirror and didn't tell her that her face was now pulled to one side, her eye had dropped and her nose was flexed to the right. Because the face was so distorted her tear ducts became squashed shut. This was another setback and, worried that the absence of tears would let the eye become infected, the doctors sewed her eyelids together – using bright orange stitches.

Weeks went by and her face didn't improve. Eileen was simply told that she was ill and needed to stay in bed. At no point did anyone mention her face, nor did Eileen ask anything about it. It never dawned on her that the reason she needed someone to feed her with a spoon was that her mouth was the wrong shape. One morning Eileen discovered the truth.

One day the ward was short-staffed, and there was a new nurse on duty who I had never met before. She took me to the bathroom and as I went in I was faced with a full-length mirror. I looked in the mirror and looked away. Then I looked back. I saw a monster glaring back at me. I stared in horror. By this point the nurse was busily running a bath, and I asked her, "Who's that?" It didn't register that it was me.

"It's you," replied the nurse.

I said, "It can't be." "Yes, it is," assured the nurse, and after a pause added, "Didn't you know?"

I remember screaming and yelling. I was hysterical. I never did get a bath that day!

Up until then I had been quite vain. I thought that looks would get you where you wanted to go in life, and looks said things about you. But now I suddenly understood why my little boy had fainted when he had seen me for the first time after the operations – the shock must have been horrific.

Eileen had a new face and a new life. Gone was the previously open and cheerful mid-thirties mother, and here was a recluse. "I didn't want to see anyone. I was sure my marriage would go, that my children would disown me – I mean, how could they bear to look at me?" she recounted quietly. "I shut everyone out – it was awful." The fact that she had been speaking with people, laughing and smiling for the previous month was no comfort; in reality it only served to increase her sense of anguish. "I just could not take it in that that person in the mirror was me – it changed my life completely."

They offered plastic surgery, but Eileen said she would rather die. This was balanced to an extent by her desire to be with her family. One Sunday the whole of her family showed up. "This did help, because they still treated me as if I was their sister. I am the eldest of eight children and I was the same to them. It helped for that period, but not once they had gone. While you are in hospital sat in that bed, you have a lot of time to think."

Over the following years she has had a series of operations, each helping to restore aspects of her facial appearance. "But I still wasn't happy with myself. By the time I came out of the hospital it was autumn and the leaves were just coming off the trees. There was no counselling, they simply told Alan to make sure that I did go out," she said. Eileen didn't want to go out socially and she only went when Alan dragged her.

My aim was to hide myself away. I didn't want anyone to know, and I didn't want anyone new to see me. Beforehand I used to love going dancing and things, but now? Now I wouldn't allow myself to think about going out – it was impossible. I didn't want Alan to be associated with me, because I felt that he was going to introduce me to people as his wife, and they would think, my God, fancy living with that. I felt utterly worthless, useless and ugly; no use to anyone. I felt terrible for Alan and worried about what the children's school-friends would say.

In reality, the children coped remarkably well. Even so, Eileen found that she couldn't cuddle her kids. "Who would want to be hugged by an ugly beast?" she asked. "When I was young I always wanted to be the centre of attention, but not like this. I went to a wedding, but when it came to the photograph I looked away. I didn't want to spoil their photos."

It wasn't just Eileen's appearance that kept her at home. The operation to remove the tumour had also affected her balance.

Add to this months of lying in bed and she needed to learn how to walk again. Even if she had wanted to, she wouldn't have been able to go out on her own.

Eventually, after more than 25 years with her left eye stitched closed, one surgeon took the plunge and reopened it. "It was amazing to have that part of my face restored. I thought this surgeon was the most wonderful man I'd ever known. My family kept looking at me in amazement because I had my eye back."

Eileen can now look back on the darker days of the previous 30 years and laugh. "I had this thing, this idea, that if I kept my lips together then no one would notice – but instead I just looked miserable," she laughed, "I'm happier now – I've come a long way."

> If someone offered me the chance of a complete face transplant to give me a totally normal face, I'd refuse it. I am happy with my face. I've learnt to cope and it is me. The Changing Faces charity has taught me that it wasn't the way I looked, it was my personality that mattered. I am me, whatever I look like. My looks are other people's problems, not mine.

Eileen's story graphically shows a person's reliance on their appearance as part of the way that they build their self-esteem. Take this positive self-image away and the person is left struggling. Their interaction with other people and their outlook on life is fundamentally shifted. In Eileen's case her attitude moved from one that was positive and outgoing, to one that was uncertain and fearful. It's equally interesting to see how she has recovered from this position. First of all, the art and science of plastic surgery has been helpful in removing some of the extreme elements of the way her face changed, but equally important has been the counselling. This again has changed her outlook. The unusual nature of her face is now something that she needs to help other people come to terms with. It's not a problem for her to fight. Once more her character has shifted a little.

"I have changed," she says firmly.

I am more compassionate towards other people. I have changed my whole outlook on life. A lot of the old me is still there, but it is a different me. I have more patience with people with disabilities. I never had much time for people who were less capable. I feel for them now. I know what they are feeling.

It seems indisputable that our physical characteristics, our bodies, can be assets that we exploit for benefit, characteristics that we learn to live with or features that get us into trouble. They can assist us form relationships with others, or cause us to hide. But, whatever their influence, an essential component of *being me* is to live within my body.

a conscious being

Given that philosophers have spent centuries debating what it means to be conscious, it may come as a surprise that we don't seem much closer to forming an agreed position. The best we can say is that most, but not all, agree that consciousness exists and is a critical part of what it is to be human.

It was in the seventeenth century that the words "conscious" and "consciousness" first made an appearance in the English language. Both terms draw on two Latin words, *con* meaning "with" and *scio*, meaning "I know". Put together, consciousness is the ability to know something and to share that knowledge. You might share it with yourself and be self-conscious, or share it with others. Initially the word was strongly linked to the knowledge of wrongdoing, but soon it mellowed and became the opposite of ideas like dormancy, dreamless sleep, swoon and insensibility.[1] Edinburgh-based consultant neurologist Adam Zemen says that we use the term for three slightly different shades of meaning. There is conscious as "awake", conscious as "aware" and consciousness as "mind".[2]

Later in the chapter we will meet Anna Putt, for whom consciousness arguably saved her life. At a point when the rest of her body had failed, relatives suddenly became aware that she was conscious and members of the medical community redoubled their efforts to keep her alive. It might be difficult to define, but for Anna, consciousness became a compelling sign of life.

Before this, it is worth having a brief look at a few of the more prominent attempts that have been made to get a grip on consciousness. One of the insuperable problems of making sense of this facet of ourselves is that it is our ability to be conscious that has to investigate the issue. The problem is a little like getting a ruler to measure itself – it's easy to come up with any idea you fancy and challenge people to disagree. All the same,

[1] Davidson W. L. (1886), *Logic of Definition*, cited in the *Oxford English Dictionary* entry on "consciousness".

[2] Zemen A. (2002), *Consciousness: A User's Guide*, Yale University Press, pp. 16–21.

some people have made better attempts at unravelling it than others.

To be ...

In ancient Greece, philosopher Aristotle (384–322 BC) developed a scheme for how he thought the soul operated. In modern terminology we can equate much of his notion of soul with our quest for the basis of consciousness. Aristotle had been a student of Plato (428–348 BC), who was himself a pupil of Socrates (469–399 BC), and as such became one of the most respected thinkers of all time. Aristotle divided the soul into two broad categories, rational and irrational, each of which was made of two independent parts. The rational component was subdivided into a scientific section, which was great at absorbing facts, and a calculative section that could manipulate these data. The irrational half was subdivided into desiderative and vegetative parts which together drive instincts and behaviour.[3] For Aristotle, consciousness would be generated more from the rational half of the soul, but he would never have considered trying to narrow down where that existed in the body. He was more interested in concepts.

Thinking didn't stop with Aristotle, and as BC rolled over into AD a few people had become more interested in the physical nature of life and started to pay more attention to what bits of the body did what. The philosopher and medic Claudius Galen (AD *c*.130–*c*.201) developed a strategy based on his assumption that blood was the critical part of the body. He had been born in what is now Western Turkey, but found himself tending wounded gladiators in Rome. This brought him face to face with blood in a big way. He came to the conclusion that there were types of blood. Blue, slow-moving blood travelled through veins. This was a fluid that moved from the liver to the body, distributing nutrients as it

[3] Vardy P. and Grosch P. (1994) *The Puzzle of Ethics*, London: HarperCollins-Religious, p. 24.

went. Red blood, on the other hand, was more interesting. This liquid had been refined in the fiery furnace of the heart's chambers; a conditioning that imprinted vital life into the blood. The arteries distributed this life-giving liquid throughout the body, passing some to the brain. Here it was further refined, and imbued with animal spirit, before migrating into the rest of the body via the nerves. This animal spirit was, he said, unique to humans, and was the stuff that made us the conscious people that we are.[4] For Galen, blood was the seat of consciousness, though what consciousness actually was lay unresolved.

While we would now see these ideas as quaint, Galen was heading along in the right direction, because he identified the porridgy mush of material inside our heads as something that played a fundamental role in our intellectual and rational being. The science of the nineteenth and twentieth centuries has now shown why his inklings were right, because we now see a tight correlation between the brain and all conscious activity, albeit in a way rather different from the manner envisaged by Galen.

But before we consider recent neuroscience, it is worth pausing in the seventeenth century. Here we find that French philosopher René Descartes (1596–1650) was taking a different tack in his quest to sort out the fundamental nature of a human being. As he worked on the issue, he became increasingly convinced that everything was open to doubt. In this world of uncertainty he looked for a starting point that would be stable. Eventually he decided – he was sure that he could not doubt the existence of his thoughts. As far as he could see, you can debate whether someone else sees the same thing as you when they look at a sunset, or whether what you perceive as red and call red is the same experience that they are reporting. You can even question whether the sunset exists, or is a figment of your imagination, and

---

[4] You can read more about the philosophy of blood in Pete Moore's previous book, *Blood and Justice: The Seventeenth-Century Parisian Doctor Who Made Blood Transfusion History* (2002), Chichester:Wiley.

indeed Descartes pursued the idea until he concluded that he wasn't even certain of his own physical existence.

This left only one thing. Having doubted everything else, he was certain of one thing – he was thinking. For Descartes, it became the one certainty in a very uncertain world. *Cogito ergo sum* – I think, therefore I am. This process of systematic doubt became the central concept behind much of Descartes' work. He decided that his own thoughts were the one thing that he could rely on and use to build his understanding of everything else. His consciousness was at the very core of his being and the only thing he could trust.

Using this facet of his existence he developed *Cartesian dualism*, an understanding of the way that human beings worked that divided us into two elements. On the one hand we are made of *matter*, and this includes the material that creates the physical brain. Matter has particular properties in that you can analyse it, weigh it, measure it and divide it into portions. On the other hand there is *mind*, the ethereal and indivisible element that exists within the brain and enables thought and consciousness. In popular language, mind was the "ghost in the machine". In more prosaic language it can be described as a homunculus – a miniature person sitting inside one's head, just as a member of an audience sits at a cinema, "watching" a performance unfold on the screen of one's mind. It is the "you" inside "yourself". From a religious point of view Descartes was comfortable with the idea that the mind was the spiritual part of human being that lived within the material animal.

In recent years, dualism has gone out of favour both in scientific as well as most philosophical and religious circles. Some people argue that we are holistic mind/body organisms; that the two elements are inseparable. Others hold that mind is only an illusion created by a bizarrely capable material body. But still a critical question remains. How is it that we "experience" life? And the question is more than academic, because being conscious of our own existence and capable of forming complex relationships with a multitude of other elements forms the foundation of our life.

Quick thinking

The mid-twentieth century brought with it a new invention that led to fresh speculation about the nature of consciousness, thought and the mind that has led some to conclude that consciousness is no more than an illusion caused by rapid calculations. In the 1930s, English genius Alan Turing (1912–1954) came up with the idea that you could potentially build a single machine that would be capable of performing an infinitely large set of complex calculations.[5] The flexibility in the machine's performance came from its ability to "read" coded instructions. Each new set of instructions would produce a new output. Tragically he died young, probably by committing suicide, but he had lived just long enough to develop these ideas into the Colossus, a pioneering electronic calculating machine, which became the forerunner of today's computers.

The world soon became used to the idea that you could have a physical mass of wiring and components that was capable of running programs – more commonly now referred to as software. It didn't take long before computers became fast enough to perform their tasks so rapidly that they gave the illusion of thinking. Could this be a new way of seeing the brain? If so, this totally material way of viewing brain function could perform the equivalent of an academic exorcism, and once and for all drive the ghost from the machine.

The analogy had some strength. Computers work in binary. All their code is ultimately stored as a series of zeros and ones – it's like a massed bank of switches that can be either off or on. At the same time biologists were realising that nerves also work by sending information coded in a series of pulses. In effect resting nerves are transmitting the message "do nothing", with each

---

[5] Turing wrote his landmark paper in April 1936, but initially struggled to get it published because it was similar to one written by an American rival. "On computable numbers with an application to the Entscheidungsproble" was eventually published in August that year; *Proceedings of the London Mathematical Society*, Series 2, 42, pp. 230–265.

nerve impulse representing the message "switch on"; or in computer language they are either sending the message "0" or "1".

Given this apparent correlation between the way that brains and computers operate, some people started asking whether this could be a new form of dualism, with the computer being analogous to the physical brain, and the software representing the mind, or soul. For example, neuropsychologist Steven Pinker says, "Humans behave flexibly *because* they are programmed: their minds are packed with combinatorial software that can generate an unlimited set of thoughts and behaviour. Behaviour may vary across cultures, but the design of the mental programs that generate it need not vary."[6] He strongly denies any notion of dualism, but curiously his use of language points in that direction.

Pinker is a passionate campaigner for the notion that we are born with much of our nature and personality hard-wired, thus contradicting the idea that we are blank slates waiting for life, experience and parenting to shape us. But at times the discussions about consciousness that have arisen from modelling the brain on computers have caused people to suggest that we are born as "blank slates". A baby could potentially be born as a brand-new computer with its emotional and intellectual hard drive sitting empty and just waiting for experience and education to load it with programs and data. Or alternatively, we could be born with programs already in place, like a pre-loaded PC from a mail-order catalogue. It's the nature-nurture debate all over again.

In 1949, Oxford philosopher Gilbert Ryle published *The Concept of Mind*, in which he roundly attacked any notion that there was this two-part separation. He set out an argument that the mind was purely a machine and that there was no need to postulate that it was in some way inhabited by some form of "ghost". A few years later, American philosopher Hilary Putnam argued

---

[6] Pinker S. (2002), *The Blank Slate*, Penguin, pp. 40–41.

that the mind simply comprised the various computational states of the brain.

It didn't take long for this argument to be reflected back on itself. If a mind was really a superfast, biologically constructed computer, then you find yourself questioning whether a mechanically built supercomputer could also be described as intelligent. Could you go further and call it conscious, or even alive? There is an episode in *Star Trek: The Next Generation* where a court case is held to decide whether Starfleet's highly developed android, Data, can be dismantled for research, or whether he should be given the protection normally afforded to sentient beings.[7] Is he property, or person? Is he simply a machine, or is he so complex that it is right to call him a sentient being, maybe even the first member of a newly established species? In the fudged world of *Star Trek* they come to a fudged conclusion. Guess what? They save Data's "life". But in reality the question will always hang in the air so long as the computer/mind analogy holds sway.

There are certainly interesting parallels between the brain and powerful computers, and computer designers have at the very least learnt much by looking over their shoulders to see what biological scientists are discovering. For example, high-speed computers work by setting many smaller machines running simultaneously. They share out a task between a number of different "parallel processors" and then add up the sum of all their outputs. Twenty-first-century neuroscience is increasingly realising that you could equate each neuron in the brain with a processor. Each is a living cell that is performing a vastly more complex job than simply acting as a dumb switch. With the brain consisting of a network of one hundred billion neurons, this gives a phenomenal capability for parallel processing.

The analogy works as far as giving inspiration is concerned. But look at the detail for a moment and you soon see how far

---

[7] "The measure of a man", *Star Trek: The Next Generation*, first broadcast in the United States 26 February 1989.

computers are from emulating anything as complex as the brain. Each neuron is in itself a living cell. It grows, takes in nutrients, handles internal energy systems and makes appropriate use of its DNA database, and all that before it even starts to interact with its environment. It strikes me that we assume the brain is like a computer because they can both calculate things rapidly. This would be a little like saying that a centripetal washing-machine pump is like a heart because they both pump fluids. Take them to pieces, though, and you will find substantial differences in the way that they perform their tasks. In reality, we are decades from producing a computer that can in any way match the skills of a single neuron. The only people who suggest that we are anywhere near the verge of producing mechanised life in all its wonder are those who have never bothered to take a look down a microscope.

The claims seem plainly audacious, but reminiscent of many of the exaggerations printed around the start of the current millennium. Ray Kurzweil, an executive officer of Kurzweil Technologies, gave a classic example when he claimed in 1999 that "by the third decade of the twenty-first century we will be in a position to create complete, detailed maps of the computationally relevant features of the human brain and to re-create these designs in advanced neural computers". And the result of all this effort? "... there will no longer be a clear distinction between human and machine". As he continues he gets more excited.

Suppose we scan someone's brain and reinstate the resulting "mind file" into a suitable computing medium. Will the entity that emerges from such an operation be conscious? At what point do we consider an entity to be conscious, to be self aware, to have free will? How do we distinguish a process that is conscious from one that just acts as if it is conscious? If the entity is very convincing when it says, "I'm lonely, please keep me company," does that settle the issue?

Many people would say that a clever-talking machine is still a clever-talking machine, but Kurzweil is convinced that they will "get mad" if we refuse to accept that they are conscious.[8]

These are bold claims, and even if they were restricted to questions of intellect and consciousness I think they are highly questionable. They may help sell newspaper articles, books or shares in your company, but they are too much. It is inconceivable that a machine would ever have a biography to set down on paper that is reminiscent of any one of the stories that fall between the covers of this book. Machines might be capable, but they will not be conscious. I am not in the least bit worried that we will succeed in blurring the boundary. Comparing machine capability and human consciousness might make interesting bar-room speculation, and it may throw up some interesting ideas of ways of addressing the human condition, but it is bizarre to suggest that the two are, or ever will be, equivalent.

The "I" of the storm

The scientific study of how the brain works is, on its own, creating new views of human consciousness, and according to researchers such as Pinker, it too is driving the ghost from the machine.

> Cognitive neuroscience is showing that the self too, is just another network of brain systems ... Cognitive neuro-scientists have not only exorcised the ghost but have shown that the brain does not even have a part that does exactly what the ghost is supposed to do: review all the facts and make a decision for the rest of the brain to carry out. Each of us *feels* that there is a single "I" in control. But that is an illusion that the brain works hard to produce.[9]

[8] Kurzweil R. (1999), "The coming merging of mind and machine", *Scientific American Presents Your Bionic Future*, 10, pp. 56–60.

[9] Pinker, *The Blank Slate*, pp. 42–43.

As an illustration of how this illusion of an "I" within has been destroyed, Pinker along with many philosophers and neuroscientists points to Phineas Gage, a nineteenth-century railway worker who had an accident at 4.30pm on Wednesday 13 September 1848, while at work. At the time he was 25 years old. As the local paper, the *Boston Post*, reported:

> Horrible accident – As Phineas P. Gage, a foreman on the railroad in Cavendish, was yesterday engaged in tamkin for a blast, the powder exploded, carrying an iron instrument through his head an inch and a fourth in [diameter], and three feet and eight inches in length, which he was using at the time. The iron entered on the side of his face, shattering the upper jaw, and passing back of the left eye, and out at the top of his head. The most singular circumstance connected with this melancholy affair is, that he was alive at two o'clock this afternoon, and in full possession of his reason, and free from pain.[10]

Unsurprisingly he was severely unwell for a few weeks, and was treated by the careful if somewhat amateurish attention of John Martyn Harlow, a newly qualified local doctor. "The parts of the brain that looked good for something I put back. Those that were too badly injured and looked as though they would be no good, I threw away. I kept the wound clear, sewed it up, and Gage got well," recalled Harlow when asked about the treatment.[11] A few months later Gage felt well enough to go back to work, but his employer refused to have him back. He might have been physically back on his feet, but Gage was no longer the amiable man he had been before the accident. Instead he was "fitful, capricious, impatient of advice, obstinate, and lacking in deference to his fellows". As one of his friends said, he was "no

[10] Macmillan M. (2002), *An Odd Kind of Fame: Stories of Phineas Gage*. Cambridge, Massachusetts: The MIT Press, p. 12.

[11] Macmillan, *An Odd Kind of Fame*, p. 51.

longer Gage".[12] By the time he died eleven-and-a-half years later, Gage was having frequent massive epileptic fits and had been reduced to sitting on the street begging.

In the years following his death, Gage has gained a fame that would have shocked him. He has become a case study in many a university psychology course and an example of someone whose personality changed after brain damage. Neuroscientists now think that the explanation lies in the fact that the spike smashed the bits of the machine that normally regulate interpersonal skills.

And why is this story brought out? Because it is an example of a person whose *self* changed because the mind-machine was damaged. The idea of a solid self sitting somewhere inside was apparently routed. But was it? An alternative interpretation would be that his *self* remained the same, but that Gage was incapable of performing some of the tasks that he could do before. It would be like a violinist losing her left hand. Her musical self would not be changed, but her ability to perform would be destroyed. The case highlights the issue that you can so often interpret data to fit in with the opinion you want to hold.

Another approach to identifying the "I" within can be gleaned from the writing and thinking of neuropsychologist Kenan Malik. While reviewing the arguments for the theory of mind, he homes in on the human being's remarkable ability to make subjective judgements. We are, he says, not limited to purely objective statements. He explained the difference when I met him in the café at the National Gallery in London. "If I say Paris is the capital of France that is an objective statement, if I say Paris is a beautiful city that is a subjective statement." Human beings, he said, interact with their world not just by acting on purely physical assessments, but also by adding a value judgement to their observations.

One can also understand the distinction between subjectivity and objectivity on an ontological level, a level of being; so

---

[12] Macmillan, *An Odd Kind of Fame*, p. 13.

the planet Venus will exist whether or not humans exist, it has an objective state. Thoughts, feelings, tickles, and so on, only exist because humans exist. If we weren't around there wouldn't be beings with thoughts, feelings, pain and so on.

Malik therefore divides things into objects that exist and can be measured and ideas that are generated in our heads as we interact with other people and the world around us.

None of this would have any meaning if a person were unconscious – if they had no consciousness. It is one of the most obviously important facets of our existence, though equally it is one of the most difficult to define. Having been trained as a scientist I expect to find scientific answers to all questions of life, the universe and everything, so I initially approach this search for consciousness and its ability to generate Malik's subjectivity via a scientific mindset. It's not surprising, therefore, that the first part of this chapter has summarised a little of the scientific thinking that is going on. According to Malik, however, the approach is fundamentally flawed.

There is an argument favoured among people who think we are Darwinian genetic survival machines, people such as Richard Dawkins and Steven Pinker. The argument says that consciousness is an illusion implanted in our brains through the process of natural selection. Why is it there? Because it is useful; by which they mean it gives us a better chance of passing our genes to the next generation. We are not in that sense really conscious; we just think we are. We think that there is purpose in our lives because this illusion is useful for our survival and reproduction.

But that argument says that natural selection implants untruths into our minds because it is useful for our survival. If that is so, if all we are is Darwinian creatures and nothing more, then how do we know anything we believe says something real about the world? How do we know that science says something real about the world? So long as we believe we are simply Darwinian creatures then we cannot have any faith in what science tells us. In other words the arguments that we are animals and nothing but is self-defeating because if we were "nothing but animals" we wouldn't know that we "were nothing but animals!"

Malik argues that the only way out of this trap is to rise above our own evolution, to transcend what is physically explicable. In his language, it is to be subjective beings. But this subjectivity demands that we are conscious.

Despite the attempt to run as far away from Descartes as possible, it seems we have bumped into him once again. Not in the pure form that brain and mind are two separate elements capable of independent existence, but in the realisation that a purely material explanation of our mind is not enough.

The material of consciousness

Having said that, the material element of the brain is clearly critically important in determining how consciousness occurs. Despite its importance, relatively few eminent scientists have set out on a serious search to find how the nerve cells in the brain create an environment where consciousness can occur.

Few, but there is one notable exception. The names of James Watson and Francis Crick have become permanently linked to the discovery of the structure of DNA. Watson has continued working in the area, and became chief of one of the most eminent genetics research bases in the USA. Initially, Crick also carried on studying DNA and the cellular machinery that interacts with it. As such he revealed many of the basic ground rules that enable DNA to operate. In the last few decades, however, he has switched topic, moving his gaze to what he considers as the next major milestone in science – gaining an understanding of consciousness. Where previously few scientists feared to tread, fearing ridicule from colleagues, Crick's move to this area has given the scientific study of consciousness a new sense of legitimacy.

Along with his main co-worker Christof Koch, he published a paper in February 2003 describing the problem as he saw it, and their approach to solving it. "The most difficult aspect of consciousness is the so-called 'hard problem' of qualia – the redness of red, the painfulness of pain, and so on. No one has produced any plausible explanation as to how the experience of redness of red could arise from the actions of the brain. It appears fruitless to approach this problem head-on," they explained at the beginning of the paper.[13] Their solution is to start by searching for what they call the neuronal correlates of consciousness (the NCC), the biological structure that allows consciousness to occur, or in their words "the minimal set of neuronal events that gives rise to a specific aspect of a conscious percept".

Narrowing their task to conscious appreciation of vision, they have made some progress, but are still miles away from coming to any conclusive description of the neuronal processes that allow you to appreciate the meaning of the characters on this page.

When I read their paper I was surprised to find that they had not dismissed the notion of some form of self situated in the front of the brain that, as it were, looked on to the sensory information that is collected and assimilated at the rear of the brain. "The hypothesis of the homunculus is very much out of fashion these days," they acknowledge, "but this is, after all, how everyone thinks of themselves. It would be surprising if this overwhelming illusion did not reflect in some way in the general organisation of the brain."

Asking Koch for an explanation, he replied that the front of the cortex contains the planning apparatus that contemplates the various actions that the organism can engage in. These may be the near instantaneous actions of the motor cortex, or the long-term planning such as working out how you will afford to live in your retirement, that goes on in the more frontal parts of the brain. While he is not suggesting that this "homunculus" is

---

[13] Crick F and Koch C. (2003), "A framework for consciousness", *Nature Neuroscience*, 6(2), pp. 119–125.

identical to the proverbial little-man-in-the-head, it can almost have that appearance as the planning and actions in the cortex occur in response to the sensing that is done at the back of the brain, areas called the occipital, parietal and temporal lobes. "This creates the illusion that the front of cortex (roughly everything in front of the central sulcus) is 'looking' at the back," he told me.

So after more than 2,000 years of considering the issue, we are still far away from understanding how we think, and how we know that we are thinking. But we are increasingly aware that the intricate workings of the brain will supply some, but not all of the answers. "Two facts ... astonish me," says Zeman as he draws his book on consciousness to a conclusion. "Our experience is marvellously rich *and* utterly dependent on the brain. Any account of the nature of consciousness must do justice to these facts, and, perhaps, to a third intuition: that experience is useful."[14]

Dualism, he continues, respects our belief that experience is special, but does not explain how mind interacts with matter. The alternative to this, he calls it physicalism, is equally thin. This attempts to talk about our perception of the world in purely physical terms. It restricts itself to understanding of nerves and brain cells, or the language of machines and computers, and tries to create an explanation of experience in ways that leave mind out. He maintains that this is less than satisfactory, because the starting place was an attempt to understand mind and this merely looks at the machine.

"The area in which the inability of a reductionist materialist science to explain observable phenomena is most glaringly evident is the question of human consciousness," comments American political economist and author Francis Fukuyama, continuing that despite the best efforts of neuroscientists and computer technologists, "consciousness remains as stubbornly mysterious as it ever was".[15]

[14] Zemen, *Consciousness*, p. 341.

[15] Fukuyama F. (2002), *Our Posthuman Future*, Penguin, p. 166.

Viewed from inside

Despite this scientific and philosophical doubt about what exactly consciousness is, it clearly exists and is an important part of what it is to be human. One different approach to getting a glimpse of consciousness is to take a look at the writings and lives of people for whom consciousness is almost all they have left. A vivid portrayal of this situation is contained in Jean-Dominique Bauby's incredible memoir, *The Diving Bell and the Butterfly*.[16] It tells how the 43-year-old editor-in-chief of *Elle* magazine in France suffered a stroke, which severely damaged his brain stem. Several weeks later he woke to find that he was a victim of "locked-in syndrome", a condition that had left his mind functioning but his body paralysed, except that he could still blink his left eyelid.

One early record we have of this condition is in Alexander Dumas' 1884 book, *The Count of Monte Cristo* where one character, Noirtier de Villefort, survives a stroke but is left capable of communicating only by opening and closing his eyelids, and by vertical movements of his eyes. Emile Zola created a similar character in a book that he wrote in 1868. In this case the woman was "mute and paralysed, her face a death mask with two living eyes. Only these moved, rolling rapidly in their sockets."[17]

Just as with the two fictional characters, Bauby found that blinking out letters, words and sentences was the only way that he had of letting people know what he was thinking. Unlike the story-bound two, he had a computer that could interpret his blinks and enhance his ability to communicate. His view was that while he felt that his body was as lifeless and heavy as a heavy, steel diving bell, his mind was as agile and animated as a butterfly and its flights just as random. Describing how he deliberately takes flight from his confining body he says,

> My cocoon becomes less oppressive, and my mind takes
> flight like a butterfly. There is so much to do. You can

---

[16] Bauby J-D. (1997), *The Diving Bell and the Butterfly*, Knopf.

[17] Zola E. (1868), *Madeleine Férat*.

wander off in space or in time, set out for Tierra del Fuego or for King Midas's court. You can visit the woman you love, slide down beside her and stroke her still-sleeping face. You can build castles in Spain, steal the Golden Fleece, discover Atlantis, realize your childhood dreams and adult ambitions.[18]

The rest of the book is a startling tale of physical suffering illuminated and relieved by flights of fancy.

Bauby's situation is obviously unique and personal and his book shot to the top of the bestseller lists. This occurred not just because of the pathos running through his account, nor the grim fact that he died two days after publication, but because his condition seemed to strike a cord with many people. It strongly resonates with the idea that we are a mind or spirit that is in some way contained within a body. We see the remarkably inventive qualities of our imagination tied down by the physical limitations of our biological frame. But at the same time we see the reality that Bauby's consciousness is intrinsically interwoven into his physical being – his conscious mind is not free to fly away – body and mind are irrevocably interconnected.

I don't intend to recreate the insider's depth of feeling that you will discover on reading Bauby's book. But I would, at this point, like to introduce you to someone who has found herself in a remarkably similar situation. Someone whose consciousness stands as the part of her that is still working well, even though many other faculties have failed. She is very aware of the consequences of being a conscious being.

Anna had a tough start to life. She was born in January 1969, but her mother, who was about 35 at the time, died when Anna was only 11 months old. She had been a first child, and her father took over the role of universal parent. He did a good job of parenting and Anna grew to be happy and healthy, or so everyone thought. On 9 July 1994, she married Des Putt and the

[18] Bauby, *The Diving Bell and the Butterfly*, p. 13.

pair set off on honeymoon, returning to stay with her father for a few weeks while the final stages of a house purchase went through. Then on 29 August, a bank holiday Monday, at about 9.00am Anna collapsed in bed and was rushed by ambulance to the nearby John Radcliffe Hospital in Oxford. For at least 36 hours no one had any explanation for what had happened.

"I remember waking up that morning with a violent headache," recalls Anna in one of a series of e-mails that she sent me from her Oxfordshire home.

> I could hear my husband phoning for a doctor. This worried me because I knew we'd just moved out of our previous catchment area. Something told me that I needed an ambulance. It was very strange and frightening. My brain knew what it wanted to say, but my voice and mouth wouldn't work. It was like my brain had been separated from my body. I still seemed to be Anna Putt inside, yet it was like my brain/mind had been transplanted inside a different, lifeless body. I couldn't ask for an explanation as my voice and mouth were no longer working.

By the time Anna got to hospital she was unconscious, or at least, there was no way that doctors or her relatives could detect any signs of consciousness. "After a huge number of scans, the doctors discovered a myxoma on her heart," explained Anna's father, David Chamberlain. Myxomas are a type of tumour that grows inside the lining of the heart. They can become unstable and if they break away they will jam in one of the arteries and block the blood flow. This can spell trouble if the artery is running to the person's brain. Blocking it starves the area of brain normally supplied by this artery of oxygen and nutrients, and the nerve cells soon give up working and die. The person has had a stroke. Myxomas are hereditary, and Chamberlain now thinks it was probably a myxoma that was responsible for his wife's death. Her death certificate simply records a "pulmonary embolism", a blockage in an artery in the lung. At the time of her death there was no explanation for why the embolism had occurred – but now Chamberlain had a good idea what had caused the blockage. At the time of his wife's death few people knew anything about myxomas. They are very rare with only 10 in 19 million people having them, and they were discovered in the USA only around 1957.

As soon as the growths were detected, heart surgeon Stephen Westaby operated to remove all of the remaining myxoma tissue from Anna's heart. In comparison with many open-heart operations, the surgery was a straightforward operation, but the brain damage had been done. She was totally dependent on life support over the first days. As far as anyone could tell, Anna was completely unconscious and had little hope of recovery. The blockage had occurred in an artery feeding the brain stem, the part of the brain that organises a lot of automatic processes, such as breathing and the speed that your heart beats, and through which all communications about how and when to move the body's muscles has to pass. A large part of this area was effectively destroyed.

"We were encouraged to talk to her – one particular nurse was good at doing this, always explaining to her, 'Anna, I am just going to do this to you,'" explained Chamberlain, "although there was not a flicker of any response. Des and I went every evening after work – we always said 'goodbye' when we left."

But the blankly staring Anna was not as remote as they feared. "Somebody must have put earphones on me playing my favourite, local radio station. They were talking about how the IRA had just announced a full ceasefire. I remember thinking, 'When I see my husband later we can talk about that'," comments Anna. You see, while her relatives, friends and carers thought that Anna was in a coma, and therefore not capable of any thought, Anna knew otherwise. "I had my brain stem stroke on 29 August 1994. I was shocked when I saw a telly review of that year. This ceasefire was in the news around that time."

Anna can also remember hearing commentary coming from a helicopter that was flying around her home village with a reporter describing a huge traffic hold-up. "I remember thinking,

'I hope Des isn't late for work','' she said, or rather, she "typed" in another e-mail. Almost a decade later, Anna can remember details of what it was like to be conscious, but have no way of letting anyone know.

> To the outside world I was in a deep coma, but some of my hearing and my mind was carrying on normally. My visitors had been encouraged to talk to me. I don't remember any of their conversations at this early stage, but they definitely played tapes of Bryan Adams, my favourite music. My thoughts were as if it was a normal day.

Normal that was until she started to take in, though not understand, things that were happening around her.

> Then the first part of this strange new life became very, very frightening. There were instances where I could see, but it was always complete strangers around me. In particular I remember this persistent, intermittent beep. I felt so cold from inside out, even though I was covered in warm blankets. It was like my body had been frozen and strapped down. I couldn't understand it. The sky outside was azure blue. It looked like I was in a laboratory. An olive-skinned man, dressed in lightish green top and trousers, was trying to put a fine tube into the veins on my hand. It hurt; he seemed to be digging about. I couldn't move a muscle or make a sound, although I was screaming inside. I seemed to have no reflexes. He didn't talk; it was like he and I were in different worlds. I decided I was semi-dead and being experimented on. Where was God and heaven? ... I don't remember any more of this occasion.

On another occasion, Anna remembers being in a room where there were no "beeps", and two nurses were changing the bottom sheet on her bed. To do this they had to roll her side-to-side, but as they rolled her, Anna felt her right kneecap dislocate. She was used to the feeling because it had happened often since she was about 14. But what she wasn't used to was the patella being left out of place. She was in pain, but incapable of letting anyone know. "One nurse said, 'She looks dead', which did not do my morale any good at all," Anna comments. "I think I was unconscious when the hospital staff discovered my knee and put the patella back in place again, because I don't remember that."

One night Anna was sweating and a nurse she had never seen before changed the sheets on her bed. "She stroked my head and said, 'I'm sorry I can't do anything more for you.'" It could have cheered her up to have been spoken to kindly, but unfortunately Anna took that comforting comment as indicating that she was dying, not that there was nothing more she could do to make her more comfortable.

No one knew that Anna was spending some of her day alert and conscious. "I always look on the bright side," said her father, "We hoped that some of our remarks and the music might be getting through to her, but all we had was hope – we had no evidence to support or encourage us."

On Monday 5 September, a week after the start of this crisis, Anna's uncle was visiting. It was nine o'clock in the evening and suddenly Anna opened her eyes. "I don't remember seeing anyone I recognised from the life I once knew for ages until I opened my eyes one day," recalls Anna.

The room was brightly lit and my bed was white. The room was quiet. My uncle was standing over me, talking. I must have made some facial reaction because he moved his finger around in the air, from side to side and said, "Look, she's awake! Follow my finger, Anna!" This I did to the best of my ability. Apparently, he then telephoned my husband and father, who had just got home. I was grateful that someone still loved me, but wondered why there was so much concern about my eyes.

Much later Anna found out that her relatives had been told that she was probably blind.

"Des and I had been there during the day," her father told me, "but when they phoned we rushed to the hospital just to see. Sadly by the time we got there she had gone back to sleep."

I remember waking up in, what I assume, was the middle of the night because it was very dark. There was that intermittent beep all the time and what looked like televisions partly lit up, which I now know were the heart monitors in intensive care. It felt like my neck was resting on a bowl. Someone was pouring lovely warm water over my hair and massaging and rubbing my whole scalp. Whoever it was, was washing my hair! It felt absolutely *wonderful*. The sensations must have roused me. After they had finished, the bowl was removed. I tried to hold my head up, to look round to see where on earth I was, and failed completely. My neck appeared to have lost all its strength and the back of my head felt as if it had been filled with concrete. What had happened? I couldn't make a sound and don't remember being able to cry at all, yet I was frightened.

Later I studied my environment to the best of my ability. There were curtain rails in a square way above me and beyond that a painted ceiling. The curtains were mostly open and attached to the rails. Being unable to lift or move my head and laid flat on my back, I could only look directly up and across the room high up, where there were more rails and curtains. I could see no sign of life in my line of vision, but there were plenty of voices. One voice in particular, I thought I recognised from my past. When I had started work, my line-manager, a woman, had a very loud authoritative voice at times. I knew she had been promoted to head office two years previously. She was there giving orders again, yet clearly this was not the bank! My assumption was that this lady had come to check up on me. Indeed, why wasn't I working? I had no answer – what excuse could I make?

My family were around my bed and talking among each other. One person mentioned a forthcoming funeral the next week. I whimpered, thinking it was mine. Luckily, one of my uncles realised and calmed me down, assuring me it wasn't mine.

Three days after the first eye movements, Anna was moved from intensive care to a more normal ward, which worried her husband and father. "We felt more at ease when she was in intensive care – we thought that the move was too soon," said Mr Chamberlain. "The first night on the ward was almost our lowest ebb." It got even lower, when they noticed that DNR had been written across her notes. The acronym stands for "Do Not Resuscitate", and told staff not to begin any strenuous effort to save life if her heart or breathing stopped. At the very least, it gives the impression that doctors have already given up the fight. "I think it was put on her record after she had been moved off the intensive care ward," said Mr Chamberlain.

I am not sure what the ethics of this are, but I think they should have discussed it with us. The unspoken feeling seemed to be that Anna's quality of her life would be very poor anyway – although Anna would challenge that now – so any deterioration, challenge or threat to that would cause an even worse life, which we can't endorse.

A couple of days later Anna contracted pneumonia and was on very strong antibiotics.

Apparently, some friends of mine from a local church had arranged to meet around my bedside to say prayers on the Sunday afternoon. To quote a friend, I looked "ashen grey and ghastly". I could barely open my eyes and breathing was hard work. I felt like panicking, but could sense figures around me seemingly joined together. To me they were dancing, or at least moving around the bed. There was a sense of calmness around, but I felt increasingly dizzy. I was not frightened, but needed these figures to keep still.

In fact her friends were standing still and holding hands around the bed while they prayed – they didn't move. Anna's sensation of dizzy movement was most probably due to the nurses raising the bed end so that she was in more of a sitting position to greet her friends – her senses were in such a mess that this change of angle added to the confusion in her mind.

My friend has since explained that when they came that afternoon, they stood still and held hands all around my bedside. They quietly said prayers and thoughts. However, they did not move. At the time I was not aware these "figures" were friends from church, however, they were different from the usual groups who came. I could sense their calmness and it felt important and reassuring. My pneumonia was at its worst. Indeed my husband and father were sent home early that evening because I was "critical" and the doctors needed to be around me.

"When I phoned the next morning, the worst was over," said Mr Chamberlain.

"I would like to believe, although I was very ill, I was roused by friends and their prayers. This helped both myself and the doctors fight and treat the pneumonia," Anna wrote.

And some of the staff showed that they were more used to working with people in Anna's type of situation.

There was a registrar from Rivermead Rehabilitation Centre called Imad. I can remember the first time he came to visit me at the John Radcliffe Hospital. He actually *spoke* to me. He asked me if I would like the nasal feed tube, which I hated, replaced by a tube in the stomach. I opened my eyes wide and smiled to say yes. It was the first time I remember anyone asking *me* a question about my care.

On another occasion hospital staff were not so thoughtful. Anna remembers someone putting a hard X-ray plate under her back. "It hurt against my left shoulder blade," she told me. "It felt like it was left there for ages." She then remembers being pushed into a long dark tunnel and thinking "they" were burying her alive. "I was so terrified I think my arms even spasmed. There was a whirling noise, I expected to be ground to a pulp at any time." The machine was a brain scanner, but no one had thought to tell Anna what was happening.

> Often lots of people in white coats gathered around, always talking about me and not to me. Some of it was familiar and then they drifted off, discussing things I'd never heard of before. I wanted to ask them where I was and why? What they were talking about? How *dare* they talk as if I was not there! I felt livid, but I couldn't get it across.
>
> All my dreadful thoughts and confused state could have been prevented if only the hospital staff had told me what they were doing, even if there appeared to be no response from me. Nobody even explained what had actually happened to me. It was my father who eventually told me the truth and that I wasn't dying.

Even that only happened in November.

Anna laughs at some of the ways people behaved. For instance, at one point during her stay in hospital she was in a room on her own. Each day nursing staff got her out of bed to sit in a chair for a couple of hours. They also taped the instructions for a communication system based on winks and blinks to her locker over the other side of the room. The fact that she made no use of the system seemed to confirm that she was not conscious inside. It didn't even occur to her family that, although they knew she had been short-sighted for years, she would not have been able to read the instructions because no one thought to put Anna's glasses on, or to bring the board near enough to give her a chance. "I used to turn my head to stare at the instructions but, being more than a metre away I couldn't read them. I was hardly in a position to ask for the locker to be brought nearer or for my glasses to be brought from home."

Gradually Anna spent more time being fully conscious, and she and her family started to learn to communicate via blinks and winks enabling her to make "yes" and "no" responses to questions. She also regained some ability to control movements of her head and learnt to use a specially devised "speech" board. This was made of transparent plastic and visitors held it between themselves and Anna. The board had red, blue, yellow and green triangles placed one in each corner, and then the alphabet was arranged in clusters around the board with each character in a cluster being a different colour. To spell out words Anna looked first at the relevant coloured triangle and then at the cluster containing the letter. The friend could see where Anna was looking by watching from the other side. The process was slow, but it started to give Anna some means of letting her intellect escape.

By 23 January she was medically stable enough for the medics to transfer her to the Rivermead Rehabilitation Centre in Oxford, and then two years later, just before Easter 1997, she moved back home and started rebuilding her life.

Life is very different from anything that most of us will ever experience. If you can only move your head and blink then you are totally dependent on assistance. Being dependent also means that you need to create a fairly rigid routine. On weekdays, Des leaves for work early, and then a quarter of an hour later two helpers arrive. They get Anna up and then one stays to be with Anna for the morning. Around lunchtime another carer comes and stays until around 6.30pm. Two more carers come to the house at bedtime to help get Anna into bed.

Set on its own, that would give the impression that Anna is inert. She may need a huge amount of physical support, but her mind is still very alert. For example, she has a very accurate recall of telephone or bank account numbers.

Local radio ran a campaign and Des spent an exciting evening watching the studio as in two hours they raised sponsorship for a specially adapted car that will take Anna's chair. Her father christened this blue Vauxhall Combo the "Anna-mobile", after the well-known goldfish-bowl-like Pope-mobile. Getting out is important. "She went out to a local park yesterday and phoned me to ask for a book to help identify local wildfowl," said her father. "She loves having children around. Two of her cousin's children are her godchildren and she takes enormous pleasure in other people's family life."

Anna has two wheelchairs, one of which is electric. "She can drive herself around the supermarket – all with movements of the head," explains her father.

> She gets a mixed response. The key problem is that everyone assumes that she is incapable, but when we go shopping the only things in the trolley are the ones she has asked me to put there – it's her shopping trip. There is also a tendency for people to push in front of her. At the checkout most of the operators talk to me and ignore Anna. There is one who always addresses questions to her – things like, "Do you want to use your reward card?" Everybody else asks me.
>
> Very few people have the ability to react in a normal way to Anna – Anna wants to be treated as a normal person – told, "You look good today", or "You don't look quite as good, what is the matter?" or "Isn't it a lovely day and isn't petrol expensive!"

I called in to visit Anna and Des one afternoon. Her father was also there, along with the hundreds of frogs of all shapes and sizes that Anna has collected avidly over the last couple of decades. She was clearly someone who was very much in charge of her surroundings, and Des had built some carefully crafted cabinets to house some of Anna's collection. In preparation for my coming she had baked a chocolate cake, by giving a precise series of instructions to her helper that morning. The result was excellent.

As well as directing people, Anna can control many of the features around the home. From a menu-driven box of tricks mounted on her wheelchair, she can open and close the front door, draw curtains, operate the television; indeed the box can be programmed to take charge of anything that has an infrared controller fitted to it. A nod pushes a lever on the left side of her head, setting the cursor stepping down the list. Anna waits for it to reach the desired command and then pushes the lever again to select it.

One of the most valuable pieces of equipment now is her computer and Anna took me through to the computer room that pleasantly overlooks their garden. I'd asked earlier about the purpose of a white reflective dot fixed to the bridge of her glasses and was told to wait and see. Now it made sense. A box containing a light beam and receiver monitors her head movements, allowing Anna to direct a mouse around the screen. Specially created software places a series of different keyboards and menus on the screen, enabling her to drive standard Windows-based software. Anna selects letters by holding the cursor over each character for a few seconds. Like the texting software on mobile phones, her computer constantly tries to second-guess what she is writing, so on the whole Anna has only to type the first few letters for each word. It still demands determination to hack out a paragraph. E-mail is a great boon, and gives Anna scope for contact with lots of different people – including the occasional intrusive author.

She was intrigued when I sent her an article published in *New Scientist* of another person locked within his immobile body. Hans-Peter Salzmann has Lou Gehrig's disease, which has destroyed his nerve fibres and left him locked into a paralysed body. He can't even breathe without the assistance of a mechanical ventilator. But his mind is active. In fact it is so active that medical psychologist Niels Birbaumer at the University of Tübingen in Germany has built a computer that can be operated by Hans-Peter's brain-waves. Known as the thought translation device (TTD), the machine uses two electrodes stuck to his scalp. These pick up electrical signals produced on the surface of the brain immediately before a thought or an action.

The computer allows Hans-Peter to choose a letter by displaying half of the alphabet at the bottom of the screen. The cursor starts at the top. If the letter he wants is there he "tells" the computer to move the cursor down. If not, he waits, and the second half of the alphabet appears. Now he moves the cursor. Having selected the required half alphabet, the computer now divides the letters into two further groups, and the process continues. Eventually there is only one letter left, the one he wants. It is a labour-intensive process, and choosing isn't easy. Hans-Peter says that he tells the computer to select by "building up tension with the help of certain images, like a bow being drawn or traffic lights changing from red to yellow" and allowing this tension to "explode by imagining the arrow shooting from the bow or the traffic lights changing to green".[19]

"What a wonderful communication system. Impulses from the brain! What patience!" responded Anna after reading the article.

When it was time for me to leave, I was shown the door – literally – Anna nodded her head and a whirring motor swung it open. I left Anna and Des' home that afternoon in no doubt that Anna was an intelligent and vibrant woman who enjoyed life. Apart from her lack of movement and the consequent difference that that has made to her lifestyle, she says that she feels that she, as a person, has changed very little since the stroke. She has grown up and matured like anyone else, but thinks the only difference is that while she used to worry a lot, she is now a lot less anxious. In fact her only anxiety is that she may get another myxoma in her heart and the surgeons be unable to operate to remove it. She loves life and does not want it cut short by another stroke.

A dim light within

We all have the worry that we may have an accident or illness that would make us so disabled that we couldn't communicate, but were still aware of what is going on. Some of the Victorians went to great lengths to provide coffins for themselves with alarm signals that they could operate just in case they woke up,

[19] Quoted in Spinney L. (2003) "Hear my voice", *New Scientist*, pp. 36–39, 22 February.

six feet underground. A greater understanding of medical science now makes such inventions unnecessary, but the question has now moved from the grave to the hospital ward. The issue is summed up in the question: At what point do we turn off support?

One way of looking for signs of consciousness in difficult cases might be to use a brain scanner and a series of relatively simple tests. Working at the MRC Cognition and Brain Sciences Unit in Cambridge, Adrian Owen and his team have used brain-imaging equipment to have a technological look inside. When I spoke to him, he had just completed a study of three patients. One subject was a woman who had fallen into a coma after having a severe fever. Owen explained that previous experiments in healthy people had identified areas of the brain that increased their activity when they saw photos of friends and family. They wanted to see whether the same would happen in this apparently unconscious woman.

To do this, they slid the woman into a positron emission tomograph (PET), which allowed them to work out how active each bit of brain was. While scanning her brain, they placed photographs of faces of her friends and family in front of her. These were mixed with images of scrambled faces. Analysing the data showed that the areas that they would expect to see "light up" in a healthy person, the right fusiform gyrus, showed increased activity when she saw someone she knew well, but not with the unrecognisable faces. The patient appeared to be recognising the images, even though she was incapable of making any response. A second patient they looked at had suffered from a traffic accident. She made no response when shown the photos, but the team saw clear activation of the superior temporal gyrus, an area associated with hearing, when they played her familiar music. Both of these patients went on to make an almost complete cognitive recovery.[20]

---

[20] Owen, Menon D.K. *et al* (2002) "Detecting residual cognitive function in persistent vegetative state", *Neurocase*, 8(5) pp. 394–403.

The third patient was different. She was again in a coma following a high fever, but she failed to show any responses to stimuli. In contrast to the first two, she has also never shown any signs of regaining consciousness.

The tests are technically difficult to perform, partly because the subject needs to lie very still within the PET scanner, and many people who appear not to be conscious constantly make erratic movements. All the same, they give an impression that there may be a technological way of determining whether there is a butterfly inside the diving bell.

An altered mind

The concept of a locked-in mind brings with it an overtone that the mind is in some way untouchable, that you can strip away the physical capabilities of a person and leave their mind unaltered. We have already met a potential counter to that theory in the form of Phineas Gage, whose behaviour was radically altered by the loss of a substantial section of his brain. But survival after such gross accidents is a rare event. Much more common are examples of the use of mind-altering drugs. Just as Darwinism and genetics combined to shake the foundations under ideas of human existence and identity in the nineteenth and twentieth centuries, neuroscience and neuropharmacology stand to bring a new set of challenges to the twenty-first.

If in doubt, look no further than our classrooms. Here you will find the Ritalin generation sitting peacefully at their desks and concentrating hard at their work. Ritalin is the trade name of methylphenidate, one of a group of drugs that closely resembles "speed", the substance that sold well on the streets in the 1960s. It is chemically similar to methamphetamine and cocaine and increases attention span, builds up short-term energy levels, and enhances a person's ability to focus on a task.

It has shot to fame as a prescription given predominantly to young boys who have problems sitting still in class. In their book, *Driven to Distraction*, psychiatrists Edward Hallowell and

John Ratey assert that once you have learnt to recognise attention deficit hyperactivity disorder (ADHD), "You'll see it everywhere".[21] What they are observing are children who have trouble concentrating and have "overactive motor functions" – they move around a lot. These two doctors believe that there could be anything up to 15 million Americans with the condition. And the drugs do have remarkable effects. They can almost instantly turn an unruly child into a star pupil.

The numbers of children listed with the disease seem staggering. A 1995 study of children in the mid-west of America found that more that 12 per cent of two- to four-year-old children from low socio-economic backgrounds were on some form of behaviour-altering drug.[22] A 1998 study of one population of the poorer groups of people in Michigan found that some 57 per cent of children diagnosed with ADHD under the age of four were on medication.[23]

It's not just young children who use it. These performance-enhancing drugs are increasingly commonplace in high schools and colleges where students are realising that they can boost their exam results. One doctor at the University of Wisconsin commented that "the study rooms are as good as some of the local pharmacies here".[24] A journalist on the *New York Times* told of mothers who stole tablets from their children to boost their own concentration.[25]

[21] Hallowell E.M. and Ratey J.J. (1994), *Driven to Distraction: Recognizing and Coping with Attention Deficit Disorder from Childhood through Adulthood*, New York: Simon and Schulster.

[22] Zito J.M. *et al*, (2000), "Trends in the prescribing of psychotropic medications to preschoolers", *Journal of the American Medical Association*, 283, pp. 1025–1060.

[23] Rappley M. *et al* (1999), "Diagnosis of Attention-Deficit/Hyperactivity Disorder and use of psychotropic medication in very young children", *Archives of Pediatrics and Adolescent Medicine*, 153 pp. 1039–1045.

[24] Hanchett D. (2000), "Ritalin speeds way to campuses: College kids using drug to study, party", *Boston Herald*, p. 8, 21 May.

[25] Wurtzel E. (2000), "Adventures in Ritalin", *New York Times*, p. A15, 1 April.

There is now an active political debate about the figures. Is this level of prescribing due to an epidemic of disease, or is it caused by school practices that are tailored for the most academically able pupils, but inaccessible to those perfectly normal human beings who have other skills? As Fukuyama puts it, it could be that

> ADHD isn't a disease at all but rather just the tail of the bell curve describing the distribution of perfectly normal behaviour. Young human beings, and particularly young boys, were not designed by evolution to sit around at a desk for hours at a time paying attention to a teacher, but rather to run and play and do other physically active things. The fact that we increasingly demand that they sit still in classrooms, or that parents and teachers have less time to spend with them on interesting tasks, is what creates the impression that there is a growing disease.[26]

Impression or not, what interests me is the effect that this drug has on people. You could do a similar study of drugs like Prozac that push people's moods in the other way. These types of drugs undoubtedly change behaviour. Ritalin and its cousins calm people down; Prozac lifts depressed people up. They alter people's moods, and change a person's conscious approach to life. In many ways you could argue that they have chemically manipulated the person's mind, which in other situations was seen as the very essence of the person within the body. Fukuyama sees pharmacological manipulation of people as the biggest area of change in the near future:

> Our ability to manipulate human behaviour is not dependent on the development of genetic engineering. Virtually everything we can anticipate being able to do through genetic engineering we will most likely be able to do much sooner through neuropharmacology.[27]

[26] Fukuyama, *Our Posthuman Future*, p. 47.

[27] Fukuyama, *Our Posthuman Future*, p. 173.

His anxiety goes to the heart of why this chapter is important in this book. You can argue for or against the benefits of mind-altering drugs, but you can't argue against the idea that altering the mind alters your perception of yourself and changes your life. Our conscious appreciation of the world and our reactions to it are a large part of who we are as human beings. As conscious beings we can choose to make sacrifices, we can choose to love other people, to value them highly and seek ways of putting their interests first, we can choose to work or just sit around and take what comes. We can choose to forgive or harbour grudges. Our conscious minds are the places where we grapple with moral choices and where subjective opinions are formed and find their expression. Consciousness is the facet of our existence that enables the artist to conceive of shapes and forms that others can recognise and respond to, and as such it is inevitably linked closely to the spiritual aspects of being human.

Both Anna Putt and Jean-Dominique Bauby reveal that, while they have lost voluntary control over most of their bodies' functions so that they are stubbornly unresponsive to conscious control, their minds are very much alive and kicking. The minds that are such important aspects of human beings may be imprisoned within their immobile bodies, but take a little time and you will soon find the conscious beings within.

CHAPTER THREE

a genetic being

Throughout the history of science, and of philosophical thought for that matter, there has been a tendency to think that the most important aspect of our being is the bit that we have just discovered. The unintended consequence of this is that we often end up assuming that the most important part of us is the bit that we least understand.

The most persistent area of science to grab headlines in recent years has been genetics. The first few years of the twenty-first century have seen a wealth of new announcements, most exciting of which was that scientists have worked out the vast majority of the three billion "letter" sequence that makes up the human genetic code.[1] They have translated the human genome from characters written in DNA to characters written on paper, or more likely stored on computer discs, because writing down three billion letters requires a stack of paperback books 150 metres high – somewhere around the height of a 50-storey tower block.[2]

There is now a huge effort aimed at turning this vast store of information into usable products. At the same time a debate that seems to have gone on for centuries has resurfaced. To what extent are we the people we are because of our physical inheritance, and to what extent do we owe our characters to the environment in which we have lived? The question points to the classic nature-nurture debate, and it played a critical role in the political climate of the world throughout the previous century as two key ideologies wrestled for supremacy.

In a simplified view of the fight we would find in the left corner the Marxist elements who maintained that all people were born alike, and that any differences in adults were purely a product of their upbringing. This left wing of politics claimed that if you gave everyone equal access to education, theoretically they would all attain equivalent results. The best way to tackle criminal

---

[1] Collins F. S., Green E. D., Guttmacher A. E. and Guyer M. S. (2003), "A vision for the future of genomics research", *Nature*, 422 pp. 1–13.

[2] Moore P. (2002), *Babel's Shadow*, Lion Publishing, p. 12.

elements in the community is to create a rich system of social security. By supporting people, the argument goes, you will give them a good environment and this will enable them to become good people.

In the right corner were those who looked for signs of inherited supremacy. These people have been classically caricatured by those in the Nazi campaign who aimed to build a super race. You can add to them the early-twentieth-century eugenicists who were keen to eliminate inferior breeding stock from within their nation by forcibly sterilising criminals and those deemed to have low intelligence. The argument here is that there is no point wasting resources on poorly performing people, because they will never get any better. Society will move forward by concentrating on the achievers, and if necessary simply containing, or in extreme cases removing, the losers. In effect they were saying that Darwinian natural selection had done a reasonable job thus far, now it was time for human beings to step in and give it a helping hand.

You can see the implications of the two ideologies clearly in terms of social policy. Take the attitudes to prison, for example. The left-wing *nurture* camp would want to spend time and money on programmes aimed at enabling inmates to reform. Prison then becomes a place of reformation. The right-wing *nature* camp holds that they were genetically bound to be in prison, so there is nothing we can do but keep them there. Prison is a place of containment and punishment.

Currently the left appears to be losing. The new science of genetics seems so awesome that innate programming has gained huge popular and political support.

The broad movement from environmentalism [nurture] to genetic determinism [nature] that has occurred in psychology over the last thirty years has foreshadowed the increasingly popular belief that people are genetically programmed to become the way they are, and therefore little can be done, in the way of changing the environment, that will make an appreciable difference in improving test scores or lowering crime rates or reducing poverty, to name several conspicuous examples,

claims journalist Lawrence Wright in his book *Twins*.[3]

Twins studies

Since at least the mid-1800s, researchers interested in untangling the relative effects of nature and nurture have turned to identical twins, and in particular to identical twins who have been separated at birth. Each pair starts as a single fertilised egg, and divides into two people within the first few days of their development as an embryo. They therefore share their genetic heritage, as well as sharing the environment of the womb. They intimately share the first nine months of their existence. What then if their lives after birth are lived in very different circumstances? Surely an assessment that looks for similarities and differences between them will be able to distinguish which facets of their characteristics come from their genetic nature, and which from their environmental nurture.

The early eugenicist Francis Galton was one of the first to realise this potential. As early as 1875 he was convinced that twins separated at birth would develop in an identical manner, much in the way that two clocks if separated keep the same time. For him, life was highly mechanical.

Since then many people have studied twins and some have set up huge registries and databases. Some of the earlier studies fell into disrepute when people started to ask searching questions about the data. Among some commentators there grew a strong sense of suspicion that the results had been fiddled,[4] but more recent studies have gained a high level of academic acceptance.

[3] Wright L. (1997), *Twins: Genes, Environment and the Mystery of Identity*, Phoenix. p. 8.

[4] Gillie O. (1976), "Crucial data was faked by eminent psychologist", *Sunday Times*, London, 24 October.

Even so, it is hard to see the dividing line between cold, calculated reporting of data and conclusions, and politically motivated hype that has a tendency to stretch the findings.

Wright closes his first chapter with: "Twins have been used to prove a point, and the point is that we don't become. We are."[5] When I first read that shortly after the book was published in 1997, all the alarm bells started. Built into it was the assumption that the researchers had set out to prove their point – it immediately made me question whether they had genuinely taken an impartial look at the data. His closing sentence of the book made me even more anxious. "There is finally no escape from being the people we were born to be."[6] The problem was that to my mind the material presented in the book failed to warrant such extreme conclusions. Yes, Wright's assessment of the twins studies showed that identical twins shared much of their character, but they were still two people, often with two very separate stories. As far as I could see, they were far from pre-programmed automata who had no escape from a genetically determined fate.

But I was still intrigued by the phenomenon, and determined to track down a pair of twins who had been separated at birth to see for myself what they were like. I had no pretence of conducting an academically rigorous study to find out the exact percentage contribution of nature and nurture, but was curious to see how genetic predestination worked out in practice.

Twins in South Wales

The train got me as far as Abergavenny and a short taxi ride wound through the small main road into the South Wales valleys. The hills, clothed with the first flush of spring, begged to be climbed. They almost glowed in the low sunshine. My destination was the town of Crickhowell, just a few miles over the border from England, and home of Ann Jeremiah. Her sister,

---

[5] Wright, *Twins*, p. 9.

[6] Wright, *Twins*, p. 139.

her identical twin sister, Judy Tabbott, lives just over the hills in the next-door valley.

"We have never lived together," explained Judy as the three of us sat eating welsh-cakes and scones. "When we were born in 1946 we were the sixth and seventh children of William and Gladys Thorn. My brothers were 14, 13, 12 and seven, and there was a two-year-old sister who had pneumonia at the time and everyone was very concerned about her."

I was immediately struck that Judy had said "my brothers" – they were identical twins. Surely she should have said "our brothers"? I listened on.

Ann, they told me, came out first. They know the order because the times of each birth are recorded on their birth certificates. "With multiple births in England and Wales, the time of birth is given on the birth certificate. The time isn't recorded on certificates of singleton deliveries," explained Ann. "For some people, finding the time on their birth certificate and then realising the significance of it has been the point when they first discovered that they were born a twin."

For Ann and Judy the existence of each other was never kept a secret, even though by their second day of life outside the womb they were living apart. Because their older sister was ill, they were both "farmed out" to two of William's sisters. Ann was given to Blodwyn who was married to Jack Skinner. They were both around 50 years old and had no children. Blodwyn and Jack took Ann eagerly and soon she became part of their family. The sister that Judy was given to already had three children and was much less keen to keep her indefinitely, so after three months Judy went back to the family home; to William and Gladys Thorn, her birth parents. Ann never returned. Consequently, Ann was raised as an only child with elderly parents, and Judy was raised as one of six.

"And our homes were totally different," continued Judy, adding seamlessly to the story.

Our family home was a very small, two-up, two-down collier's house with the toilet across the road and down at the bottom of the garden. My father had bought the house – he drove the engines in the local steelworks. But Ann was brought up in a 1930s semi-detached house with three bedrooms and a bathroom upstairs, and even an upstairs toilet. So our upbringings were very different. We knew of each other's existence but we seldom met each other.

"I was never adopted," explained Ann.

My nurturing parents were always afraid that I might be taken back, so we kept our distance. There was no talking to that side of the family. There was very much a cut-off. Judy was part of her parents', our biological parents' extended family, which of course I never saw. I never met them. There was very much a schism caused because my parents never thought they were safe.

"My parents had all those children," said Judy, "that I don't think they really wanted you back. In any case, there was nowhere to put you." They both laugh in unison.

In fact the web of relationships was complex because while William was the biological father of both girls, and Judy's social father, he in effect became Ann's aloof uncle. To all intent and purposes, the twins were raised as distant cousins.

Transcribing the interview doesn't quite do justice to the way the conversation flowed, because as Ann and Judy sat side by side on two separate armchairs, they were quite capable of carrying on the same conversation simultaneously. This doesn't mean that they waited for each other to finish and then joined in, but that they both spoke over each other, much like a choir singing in harmony. They seemed to be quite aware of where the other was heading, and while using different words got to the same place in the story at the same time. On occasions one would allow the other a complete unaccompanied sentence, but the moments were rare, particularly if they got excited.

"There was one occasion when I was taken over to Judy's house to sleep the night when I was seven because my grandfather was very ill," recalled Ann.

I thought I was going over to play for the day and didn't realise that the plan was for me to stay. Then my pyjamas appeared. I was devastated. There was this tiny living-room heated by a dirty old coal fire, lit only by small windows and occupied by four big brothers; there was coal everywhere. You went up some stone stairs and walked through the bedroom where the four boys slept to get into the bedroom where the two girls slept. I was taken up, and they said, "well, you can sleep against the wall", and I can remember putting my hand against it and feeling it running with water. And remember, the loo's on the other side of the road – miles away. I screamed until they gave up and took me home.

"This for me was normal – unless you see different you don't know anything else," commented Judy. "I never stayed at Ann's."

"And that was the only time they tried it with me – I never stayed there again," added Ann, laughing.

Indeed there was little in the way that they were brought up that encouraged the two girls to feel any sense of attachment or belonging. There was nothing done to assist them to even form a friendship.

"If we had gone to the same school our lives would have worked out very differently," comments Ann. "We would have known each other, we would have been in the same class, we would have grown up as friends, we would have had history; we would have gravitated towards one another in the evenings."

But none of that happened. Although they both lived in Ebbw Vale, a town of 30,000 people, the twins were in different catchment areas for schools. Judy's area was poor, consisting of rows and rows of poor terraces. Ann's was not really middle class, but had the trappings of increasing wealth. Therefore the school was better and the expectation in Ann's home and the homes of other school-friends was high. Not so Judy. "I was the last of six," said Judy, with a resigned shrug of her shoulders. "I just tagged along. Nobody pulled you up, because it wasn't important. I mean, we didn't have books in our house. The only book I ever saw was when my older brother came back from the army and was reading a cowboy book. You could never find a pen."

As an only child, Ann also had the loving attention of two parents who were keen to see her succeed in school. She was always going to the library. "The disadvantage is that I had no idea how to socialise," said Ann, adding that of the two of them she is sure that she is now the colder and harder. "I'm sure that comes down from our upbringing. I mean, I still curl up and rock myself to sleep in bed."

Ann also had a few problems working out her place in the world. At a time where few people came from homes where parents weren't married with a common surname, she found herself with two. "I was brought up by Mr and Mrs Skinner, so the neighbours called me Ann Skinner. But at school I had my legal name, which was Ann Thorn," she explained. "When my friends came home, I would say, now remember my mother is not Mrs Thorn, it is Mrs Skinner. I always had that problem – who the hell am I."

The two of them relaxed back into their chairs, gently crossing their legs at their ankles and resting their arms easily on either side of their chair. It looked as if they had been practising the routine for years.

At the age of 10, they took their 11-plus exams. A peculiarity of the system was that different schools sent different proportions of their children to the grammar school. The method used was curious, but it meant that because Judy came from a poorer achieving area, she failed, while Ann's more affluent school gave her the environment that helped her to pass. The chances are that they had both achieved similar scores in the IQ test, although Judy would not have had a hope on the general knowledge questions. It was the area they lived combined with the lack of interest in education within her family that made the difference.

"I was devastated, do you know?" said Judy. "I can remember it now. I thought I was going. I was always in the top of the class. Maths, English, I was always at the top."

"She went to the secondary modern school, and came first every year," added Ann.

"Every year," Judy confirmed.

"Other children were transferred to the grammar school, but she never was."

"Never was, see, no body pushing," echoed Judy. "I'd bring my report home and leave it on the mantelpiece, and it would never even be opened."

"Tragic, it was, tragic," commented Ann.

"Not interested."

"I'd bring mine home from school, and my father would say; 'How you come second, then? Why not first?'" added Ann. "What a difference!"

Aged 15, Judy took a final exam before leaving school. She got 100 per cent in maths and so was eventually offered a place in the grammar school. "They had to take notice then," says Judy. Even then the twins landed up in different classes as Judy was put down a year to give her a chance of catching up. But without encouragement from home it was a failure. "Dad might have been more encouraging if he hadn't been working all the time, he was an intelligent man, was Dad," Judy added. "He worked 16-hour shifts – didn't bring the money in the house, mind, he bet it. Mum never saw it."

"My parents were much more proper – my father never even went round the pub – not even when they were older," commented Ann.

"Different culture, eh," Judy chipped in.

They both left school aged 16 – Ann with some O-level exams, Judy with nothing to show for her intelligence and efforts. Judy got a job nursing, and Ann took a job in the local offices. Up until this point it is difficult to see any similarities. Their achievements were poles apart, and their lives were set off along very different paths. Given that they had not been brought up together, and had hardly seen each other there would appear to be nothing to keep their lives running in parallel.

This makes it all the more curious that, despite their separate upbringings and situations, both girls soon found themselves in relationships with violent men, and come the age of 16, both were married and pregnant. They gave birth within two months of each other. They were each aware of the other's situation, but in no way were they socially involved.

Domestic violence often revolves around themes of power, control and choice – or lack of it. "I stayed married for 17 years, off and on, being battered. Because ... no way out – no way out. I have three children, but one son died," added Judy quietly.

"I didn't stay married long," said Ann. "I had those O-levels. I knew that I had other options. It makes all the difference, education. Aged 18 I went to teacher training college and took my son with me."

Judy's chance for freedom came much later.

When I was 28 I took a college course that the government paid me for. At last I had an income – my own money. I took every exam I could. Most people on the course took six; I took 14 and passed the lot. I thought that this was my one opportunity. I got my O-levels. I got a job in the civil service when I was 33 and I said goodbye to my husband when I was 34. I remarried when I was 42."

"I've been married to my current husband, David, for 30 years, so it all came right in the end," added Ann.

As I sat listening to this story in stereo, it was difficult to keep in mind that these two lives had occurred independently of each other. There were only moments of contact between them over four decades, surely not enough to account for the similarities, and there is no sense that either thought sufficiently highly of the other to have any thoughts of emulating her. Indeed it was only at the age of 48 that the twins really started to get to know each other. This was the point when, like thousands of other twins, they were invited to travel to America and spend three weeks doing an intensive series of tests aimed at assessing their similarities and differences. They became part of the so-called "Minnesota twins" study.

"We had three weeks in Minnesota with our husbands," explained Judy.

"We'd only met Alan, your husband, for two hours at the most before hand," added Ann. "That's how little we knew each other."

"I was terrified," said Judy. At that point she was taking a lot of stick from her mother who basically hoped that the whole business had been forgotten years ago. Dragging it up and researching it was the last thing she wanted. Spending three weeks in a university environment was also more of a social challenge to Judy than Ann. "I've always felt inferior to Ann," Judy continued, to which Ann laughs in embarrassment. "You know, she talks better than I do."

By now Ann's husband David had joined in the conversation. The weeks in America were stressful for the husbands as well, because they were about to be thrown together, and they too were wondering quite how much of a "double" each other's wives would turn out to be.

It was as they arrived in America that David spotted the twins' purses. "They do their purses the same way," he laughed.

That is seriously spooky. You come up to the check-in at an airport – they have got these purses in which the world exists, but they are organised identically. Other women have bags of clutter, these two have got purses like filing cabinets; it's all there and ranked. I'm very lazy, I just walk into the airport and know I am going to get on the plane. Because it's all done, all there.

"My husband is a dear man," laughed Ann, "but when he is travelling he is in a nervous state. He looked at me. There I was with the tickets all neatly set out in order in a carefully organised wallet. And there was Judy with an identically laid-out set. We had both taken charge and were both doing the same thing."

"Well, I had to take charge with that husband of mine," laughed Judy, explaining that her husband had been led away to an interview room at immigration after hesitating so long when asked questions about his background.

"The funny thing was watching our husbands fill in the hundreds of questionnaires we were given. Alan would put a tick in a box, think about it, and then rub it out. We looked over to David and he was doing exactly the same thing," they said in unison.

"We did have a lovely time. Laughter, oh we laughed. Holding-your-stomach laughter. You'd catch each other's eye and you would see something funny, and you wouldn't have to say a word. We'd never experienced that before."

One section of the tests related to their perception of food. This bit was conducted in Boston. For a few days the two women lived in the equivalent of a hospital ward, having their entire food intake recorded, and all their waste monitored. Here their cultural background cut in.

"The menu was huge. Well, Judy chose everything. When she went to collect it she could hardly lift the tray," laughed Ann. "I asked why she had got it all – 'Well,' she replied, 'I might be hungry – I was brought up one of six, you know.' "

Boston was also Judy's first experience of an art gallery. "I was 48, and I was shaking," she said.

"I remember you seeing that Picasso, and you saying, 'I can see it, I can see what he's saying,' " added Ann.

For seven or eight weeks after they returned to Wales they meet in the hairdresser's, and gradually the relationship grew. "What is so wonderful is that our children have all got to know each other as adults and are wonderfully supportive," they said. "That has been great."

I asked them whether despite the difference in the environments that they grew up in, and the different life stories they have to tell, they felt that their characters were similar. The Minnesota trials marked them down as being remarkably similar. "The correlation was very strong," said Ann.

"If we saw something on the TV and you asked us what we thought of it, we'd come up with the same answer," replied Judy. "I might not have the fancy words, but we'd mean the same thing – I can tell Ann my innermost secrets, because I know she will feel similarly to me."

"Even though our upbringing is different we would be politically the same," added Ann.

"In terms of difference? I think that my self-esteem is lower than Ann's, but that I put up with more from other people. I had four big brothers – I had to learn to keep my head down, literally. You wouldn't pass three of them without them clip you – not to hurt you, mind."

In terms of health they were both similar, except in terms of general fitness. In the American study they had had to run on treadmills and were asked to carry on for as long as they felt comfortable. They knew that the average time that most twins spent on the treadmill was four minutes. Judy struggled to beat the four-minute mark and was completely exhausted. Ann ran on to 12 minutes and was ready to stay. The difference? Judy had smoked. Soon after coming home she joined a gym and gave up smoking. "Now I can do 12 minutes," she laughed. "Not out of a sense of competition with Ann, just for my own satisfaction."

From my viewpoint it was clear that they were both strong personalities, confident in themselves, relaxed, intelligent and open.

I asked whether they felt that the link between them had a spiritual element. Was there a connection that was more than a physical one?

"When I was married to my first husband and we had had one of our 'episodes', Ann would always turn up on the doorstep within a few days," said Judy. "I always thought that it was strange." There was also the tragic time when Judy's eight-week-old son died. It was a sudden death, nothing expected. Ann was on holiday, but on her way back she got off the bus a few stops early and went around. "This was extremely unusual, but as soon as Ann held my hand I could cope with the emotions."

"The same thing happened at my mother's funeral," added Ann. "Judy held my hand and I felt I could cope with the grief."

"Last year our mother died, aged 94, and she smoked 60 a day, and ate all the cream cakes. She was a big woman for most of her adult life. The children were all very upset, though it had much less impact on Ann."

"I was listening to them talking," said Ann. "Obviously she had been a good mother, but I realised I had no idea what it is like to have brothers and sisters. At times they have talked to me as if I was Judy. I'm a bag of nerves before I go to family events, I don't know how to relate to them. They are part of a different culture."

"But they all think of you as a sister," added Judy.

The combination of similarity and difference is played out in one of their pastimes. Both are keen card players, but even here their choice of game reflect their similar characters lived out in different cultures. Judy plays Nine Cards Don, a working-class game. Ann has always played the more middle-class game of Bridge. Both games, however, involve working with partners and "trumps" and play to similar card rules. They each require the same mental skills.

As we have seen frequently, a large part of who we are is shaped by the way other people see us and interact with us. This reaction can have peculiar twists in the case of identical twins.

"Judy came to an auction with me and a woman came and stood in front of us – she stood there and stared from one to another, it was quite unnerving," said Ann, and then added,

But a funny thing happened 15 or so years ago – Judy had gone with her boyfriend to a pub a few miles out of Crickhowell. The woman there knew David and I really well, but didn't know about Judy. She thought it was me with another man. She refused to speak to me – I couldn't understand it at all for a little while. Then you happened to say that you had called in there for a drink. That's it – so Judy, myself and David went up for a drink – problem solved!

Friends and relations are also good at spotting details of behaviour. You can see someone in the distance and without recognising the detail can tell who the person is by the way he or she moves. "When you are walking along you don't pay that much attention to the person because you assume certain things – it is only when you look you realise, because the body language that comes through to you is such that you could be with either. It is only by looking that you see the dissimilarities," said David. "I was walking past the Albert Hall in London one day and talking with Ann about the frieze on the outside, and I realised I was talking to Judy."

Intriguingly, despite their very different experience of childhood, both Ann and Judy have brought their children up in very similar ways. Both have expected almost identical standards of discipline from their children, and both have worked to let their children know that while they are important they are not the centre of the universe; there are others who need to be taken into consideration when making decisions.

"You are supposed to learn your parenting skills by referring to your own experience – they had totally different parents, totally different family circumstances and environments and came out the same," said David. Now that Ann and Judy see much more of each other, so too do their now grown-up children and the similarity of their upbringing helps them get on well.

I stood on the platform waiting for my train and reliving the interview. Clearly Judy and Ann had much in common, but equally they were their own people. In terms of what it is for each of them to "be me", their shared genetic heritage appeared to have a curious imprint, but more in the way of a watermark in expensive paper, rather than the ink used to write the letter. The similarities are there in the background, and had influenced their lives, but they are still two very distinct individuals.

In praise of genes

Ever since Austrian monk Gregor Mendel (1822–1884) studied carefully controlled plots of peas and flowers in his flowerbeds at Brno Monastery in what is now the Czech Republic, we have been aware that our physical features are passed on by physical elements. The twentieth century became the period in which we discovered the basic rules and materials buried in the cells of all living things that let this happen. As with any field of discovery, genetic research has tended to generate more questions than answers. There are probably endless questions about the precise mechanics of genetic control, but there are also the philosophical problems, at the root of which is the obvious – To what extent am I my genes?

At one level, you are 100 per cent the product of your genes. Without them you would not exist. Their instructions enable all of your organs to be in place by the time you have spent only 10 weeks in the womb, and other instructions allow you to develop and grow. The whole concept of natural selection is based on the premise that there is variation in these genes. On the trivial level, some people, whose families have lived in tropical areas for many generations, have skin tones that protect them from the damaging effect of sunlight, while others from more temperate areas have no need for this pigment. Alternatively some people will have variants of genes that give them an ability to have greater stamina than do others. If this isn't the case then we can throw out almost every modern biology textbook.

A similar line of enquiry has investigated the extent to which human genes are unique, and the extent to which they are shared by other organisms. The question has been bandied around for decades, but with increasing masses of data the picture is clearing. The best estimate is that we share the vast majority (95 per cent) of our genetic code with chimpanzees.[7] Before we

[7] Britten R. J. (2002), "Divergence between samples of chimpanzee and human DNA sequences is 5%, counting indels", *Proceedings of the National Academy of Science*, 99 (21) pp. 13633–13635. This is a reduction from Britten's previous and much publicised estimate of 99 per cent.

get too carried away and say that makes chimpanzees and humans virtually identical, we need to remember that we share around half of our genetic instructions with bananas.[8] For me the implication from this is that the message is in the detail. While there are vast volumes of genetic information that keep our cells running, and these are indeed shared throughout the living world, it is the tiny differences within commonly used genes, plus the small percentage of unique genes that give human beings their uniqueness. On top of this, minor alterations in the fine detail within the genome give individuality within the species.

The more we understand about genetics, the more it appears that the differences are less in the bulk of the instructions, and more in the tiny bits of code that act as switches. I was amused to discover recently that comparing human genetic codes with those from mice shows that buried amid the unused genes sitting in human chromosomes is a set which if activated, could enable us to build a tail.[9] The reason we don't have a tail is that we don't have the switch to turn this set of instructions on. This little difference in genetic sequence makes a huge difference in our physical appearance.

That said, we are still faced with the question of whether our character is also genetically pre-programmed. What is the involvement of our genes in determining our nature? In his book *Our Posthuman Future*, Francis Fukuyama defines human nature as "the sum of the behavior and characteristics that are typical of the human species, arising from genetic rather than environmental factors".[10] At first you could be mistaken for thinking that Fukuyama is claiming that our genes are the only things that influence our nature, and that the nurture of our environment has no influence. As his chapter continues, however, it becomes clear that this would be wrong. For Fukuyama, genes

[8] *New Scientist*, 1 July 2000, pp. 4–5.

[9] Mouse genome consortium (2002), "Initial sequencing and comparative analysis of the mouse genome", *Nature*, 420, pp. 520–562.

[10] Fukuyama F. (2002), *Our Posthuman Future*, Penguin, p. 130.

set a range of parameters within which a human being operates. Your genes may give a potential range for your height, athletic ability or even intelligence, but then your nurturing environment will play a role in determining how much of it you fulfil.

Seems reasonable, but it doesn't gain uniform agreement. Far from it. Voices like that of American neuropsychologist Steven Pinker strike a more strident note.[11] In *The Blank Slate* he sets out to ridicule the idea that we are born "blank slates", the idea that our environment and experience write on as the stories of our lives are created. The theory of "blank slates", or *tabula rasa* as they were called at the time, was developed initially by John Locke (1632–1704) in the seventeenth century. Pinker pours scorn on the work of thinkers such as B. F. Skinner and John B. Watson who have taken Locke's ideas further and created the science of behaviourism, which rules that you can mould a person's behaviour if you bring them up properly. No, says Pinker, science has now moved us to the point that we can not only rule out the effect of our immediate nurturing environment, but we can also quantify the extent to which genes and other external events influence who we are. Reaching his conclusion via an analysis of the data from twins studies, he claims that genes account for 50 per cent of our personality, our shared upbringing (i.e. our families) accounts for nothing, and the unique events that each of us encounters is attributable for the remaining 50 per cent.[12] Wright is in agreement, pointing out that the Minnesota team asserted that "none of the environmental variance is due to sharing a common family environment" and that "the effect of being reared in the same home is negligible for many psychological traits".[13]

Likewise, a subplot of Pinker's book would appear to be his desire to convince the reader that parenting is a waste of time; whatever

[11] Pinker S. (2002), *The Blank Slate*, Penguin.

[12] Pinker, *The Blank Slate*, p. 380.

[13] Wright L. (1997), *Twins: Genes, Environment and the Mystery of Identity*, Phoenix, p. 55.

you do as parents you will have no influence on your children. Children, he says repeatedly, are not blank slates for parents to write on, but they are, in the main, pre-programmed and influenced by events uniquely experienced by each person – by each child. Here Pinker seems to turn a blind eye to the obvious reality that parents have one-to-one involvement with their children, so even by his assessment parenting is not a waste of time.

Even this seems positively mild when set against views like those of genetics pioneer James Watson. Watson is keen to use genetic technology to effectively eradicate low intelligence from human society. He envisages a world where embryos are routinely screened to ensure that we give only individuals with high potential for doing well the chance of life, and expects that technology will find ways of enhancing what is naturally present in the human gene pool.

You can argue whether or not we are technically capable of administrating a genetic fix for low IQ, or even detecting and destroying embryos with low potential intelligence, but this doesn't lessen the thrust of his claim.

> We know that if we go to homeless people there is an underclass with a very strong mental disease component. Those people can't pull themselves together, the brain just won't allow it. So it is not that they are weak in character, they are seriously unequal.[14]

For Watson, genes are everything. Once you know the genes involved and can manipulate them at will, then stupidity is history. Whatever you think of the political undertones in his comments, it seems to me more than likely that this great venture will inevitably fail, because genes are only part of the story.

In the meantime, genes have been used as a lawyer's plaything. In one famous landmark trial, defendant Stephen Mobley claimed that he was genetically inclined to be violent and therefore was personally innocent even though he had killed

[14] Connor S. (2003), *Independent*, 3 February.

someone. In 1991 Mobley had shot John Collins in the back of his head as Collins cowered on his knees behind the counter of the pizza parlour where he was working. Mobley's defence team pointed to his family's history of violent and antisocial behaviour and claimed that the cause was genetic. They cited Dutch research showing that males with a defect in a particular gene, one that codes for monoamine oxidase A, tend to have mental handicap, and are violent and impulsive. Monoamine oxidase A is an enzyme that controls the levels of chemical messengers within the brain, and the theory is that without working copies of this enzyme, the brain's function is disturbed.[15] The "criminal gene" entered popular language, although the evidence for its existence is still sketchy.

One of the problems comes from the analogy whereby genes are called the blueprint of life. A blueprint is the final set of diagrams that intricately marks out the location of every component of an article. Every nut, bolt, screw, length of wire, sheet of steel, lump of plastic – they are all there, drawn to scale and in perfect position relative to one another. This expression has slipped into use when talking about genes without anyone really thinking about it and the phrase probably reflects the number of chemists and engineers who have been involved in genetic research. Biologists know little about blueprints – life is not under that scale of control.

Genes act more like a recipe. An engineer building an apple pie would have to mark out the position of each slice of apple, its weight, thickness and density. A recipe is much less precise. It simply sets out a list of ingredients, the order a cook needs to mix them in, a few general instructions about whisking, stirring or rubbing together and the temperatures they will need to use to bring about the chemical and physical changes that make the food safe and pleasant to eat. Recipes are very short and thin on

---

[15] Brunner H. G., Nelen M., Breakefield X. O., Ropers H. H. and van Oost B.A. (1993), "Abnormal behavior associated with a point mutation in the structural gene for monoamine oxidase A", *Science*, 262 pp. 578–580.

detail – blueprints go on for pages. If the genetic code were written as a blueprint it would need to be vastly larger than it currently is. It would also be massively less flexible, restricting any concept of adaptation or evolution. The whole idea of being dominated by our genes is based on a false understanding of the science of genetics.

The blueprint idea also suggests that we are "fixed" at birth – that our genes have built us, and that is the way we will be forever. I can remember having it drummed into me at school that your brain is built shortly after you are born and thereafter all that happens is that you lose brain cells. It is an image of an organ that is built once, and then slowly degenerates. The picture couldn't be further from the truth. Yes, there is progressive net loss of brain cells through life, but the cells in the brain are far from passive. The billions of nerve cells, neurones, that form our brains are shaped with a largish blob at one point, and then long tendril-like extensions that reach out and physically touch other neurones. Each cell may have tens of thousands of these delicate branches, making a total of something like 100 trillion connections within the brain. These, however, are far from stable. They are constantly reaching out in different directions, breaking old links and building new networks of connections. Why? They are responding to environmental stimuli and deliberately turning on genes that enable the remodelling. The outward impact of the process is what we call learning. In this case, the genes haven't determined the structure of the brain, they have simply enabled the brain to respond and reshape as needs must.

In a slightly less complex example, you can have all the genes you like for building powerful muscles, but if you don't exercise, the genes won't get turned on and your muscles will soon fade away. Just look at the sparrow-like appearance of a person's leg after it has been in a cast for six to ten weeks. Nothing has changed in the genetic instructions in the leg's muscles, but through a few months of poor use they will have withered away.

I guess that, like so many, I approach the subject of genetic determinism with my own set of beliefs and am not totally surprised that there was nothing in my conversation with Ann and Judy that caused me to doubt my feelings. Clearly there are similarities in the ways that Ann and Judy approach life and the way that they carry out tasks. Their body language is remarkably similar and they look sufficiently alike that on occasions they can be mistaken for each other. But equally clearly they have lived their own lives. Although each has fought against violent domination, Ann broke free quickly because her background gave her the education that gave freedom, while Judy had less options and was consequently more tied down. Ann strikes you as being more self-assured and has the courage to take on new tasks at a whim, while Judy is more cautious and in need of encouragement from others.

The problem with so much of the research in this area is that all too often it has been driven by people wanting to find evidence for their own political dogma, rather than taking an impartial view of reality. Going public with the implications of their findings has also meant that people have lost jobs and livelihoods for publishing their results. All the same, it would be bizarre to suggest that either of the two elements, genetics or the nurturing environment, had no effect on a person. Most commentators now agree that both your genetic endowment and your surroundings go to make up who you are.

Finishing a book as Wright did with the line "there is no escape from being the people we were born to be"[16] might appear to be an exciting journalistic touch, but it is clearly an overstatement of reality, and even "exciting journalism" should seek to tell the truth. Both you and your story are influenced by your genes, but not dictated by their heritage.

[16] Wright, *Twins* p. 139.

a historic being

Most people would say that their story, their life, started the day they were born. For David Barker this would be Wednesday 29 June 1938. Some people would say that it started nine months earlier on the day egg and sperm met. This is the ultimate beginning and before then, no event could have any meaningful impact on someone's life. As we will see, however, Barker has amassed credible evidence that his history started much earlier. In fact, he has uncovered evidence that his personal risk of having some forms of heart disease or diabetes is radically influenced by the diet his grandmother was eating and the circumstances she lived in while she was in the early stages of pregnancy, carrying David's mother in her womb.

Think about it for a moment to make sure you have grasped the scale of this concept. We are very used to people telling us to watch our diet and exercise if we want a healthy life. We glibly say, "You are what you eat." But now the suggestion is that we need to watch our lifestyle, not so much for ourselves, but to give our grandchildren the best chance in life. To a very real extent Barker is suggesting that his biological and physical story started at least two generations before his birth. Let's study his story, because this in itself is an example of the historic nature of a person. His own life history and chance meetings are very much at the core of his work and success.

David Barker was born in Battersea, London, but in 1940 he joined the biggest movement of children in the history of the world. Three million children left London. He moved to the country just north of the city after his house was bombed during World War II. He spent the rest of the war in the idyllic village of Much Haddam in Hertfordshire and, despite the upheaval, he was lucky to go with his mother and brother – many other children ended up being separated from loved ones.

"I have never known a time in my life when I wasn't going to be a doctor. No idea why, there was no family background – I came from a family of engineers – but I just always wanted to do that," Barker explained to me when we met in the opulent surroundings of the Royal Society of Medicine, London.

From the outset, Barker showed that he had the potential to fly high. "I got my A-Levels two years early, at 16, so the school asked me to leave because I was a nuisance," he said with a mild chuckle. "I hadn't got anything to do so I went to France for six months. I was supposed to be studying French at the University in Grenoble." In fact, Barker claims that he never went near the university, but instead pursued the interest he had developed in creepy-crawlies while studying natural history at school and headed off into the Alps to collect beetles. Working on a farm in Normandy expanded the range of his collection. Already Barker was turning into an avid collector of information.

"Then I moved into botany and went out to a small island in the north of Iceland called Grimsy to collect plants for the Natural History Museum." To an extent he worked as a modern-day Darwin, striving to fill the gaps in the vast catalogue of life on Earth. The island was chosen as it is right on the Arctic Circle and about 70 miles from land. This means that it has a limited number of plants, but many are unique to the location. Barker was learning to gather obscure data from unusual locations – it was a skill that would serve him well.

Aged 18, he returned to London to study medicine at Guy's Hospital, pausing halfway through his medical course to take a degree in anatomy. "The anatomy course was run by J. Z. Young, a massively influential figure who wrote two huge books, one called *The Life of Mammals*, the other *The Life of Vertebrates*. He was just an amazing biological polymath. And one of the things he did was embryology." Unknown to Barker, this study of the life of an embryo would give a scientific grounding to a critical element of his future career.

Having gained his medical qualifications, Barker moved to Birmingham to study with social medicine specialist Thomas McKeown. McKeown had a particular interest in fetal and placental growth. While there, Barker studied whether there was any evidence to support the idea that mental subnormality was linked to any specific events within the womb. There was a lot of concern that mental problems might be caused by babies being damaged by some unfortunate event or accident as they were being born. After three years of study, Barker's PhD thesis concluded that mental subnormality was not normally related to events around the time of delivery, but that mental deficiency has its origins much further back. It was more likely that, birthing difficulties were linked to brain damage, the bad birth came as a partial result of brain abnormalities, rather than the other way around. The research had, however, kindled in his mind the importance of life in the womb, and the concept that this may have a tangible effect on life thereafter.

A budding disease-hunter

Missing clinical practice, Barker went back to train as a consultant in internal medicine. Having jumped through the appropriate hoops over the next three years, he set off with his young family to go abroad. With a grant from the Medical Research Council, he went to Uganda for three years with the aim of trying to work out how a bacterial disease that causes skin ulcers was spread. In 1948 a team of scientists had found that these ulcers were caused by a type of micro-organism called *Mycobacterium ulcerans*. This bug comes from the same family of terrors that cause leprosy and tuberculosis.

The disease normally goes by the name of Buruli ulcer disease, and Barker examined people with it who were living in the Busoga District on the east side of the Victoria Nile, north of Lake Victoria. Prior to 1965, no one in this area had caught the disease even though there were plenty of reports that it regularly infected people in other parts of the country. There was no obvious reason to see why the two populations were so different in their vulnerability to disease. The key to unlock the previous lack of the disease and its recent outbreak was to discover how the bacteria were transmitted. Its close cousins, leprosy and tuberculosis, spread from person to person, but the fact that people had frequently travelled between disease-free and infected areas of the country without carrying the disease suggested that person-to-person transmission was unlikely for Buruli ulcer disease. There seemed to be no immediate source of the bacteria.

"I eventually had the bright idea, the fairly obvious idea, of saying well, if this is a relatively new disease, what do the local people think?" said Barker. "There were three areas in Uganda where it was occurring and I just asked people, 'What do you think causes this?' and they all said, 'Oh, it's the flooding of the Nile. Until the Nile flooded we never used to see this.'"

Between 1962 and 1964 exceptionally high rainfall caused the level of Lake Pretoria to rise and the Nile to flood. It stayed flooded and completely changed its geography. In places that had previously been fordable, it was now too deep even for elephants to cross, and a lot of swamps were created in the slower sections of the river – the very sections where this disease had suddenly emerged. "And you know, what the people were saying felt right," said Barker, who started detailed studies of exactly which people were getting the disease and what each person did. In particular he concentrated on their contact with the different water sources. It turned out that if they got their water from the swamps they got the disease and if they used boreholes they didn't get it. All the evidence eventually pointed to the idea that the disease was being harboured in the swamp vegetation. "The research was going well, but then Idi Amin turned up in 1971, and it all got extremely dangerous, so we all came home after three years."

Uganda's loss was the UK's gain. "There was a new medical school in Southampton and Sir Donald Acheson who had just founded it wrote to me and asked me to join them." Donald Acheson was a clinician and an epidemiologist and at the time there were fewer than half a dozen people who had both direct contact with patients, and had skills in chasing the root causes of mass outbreaks of disease. A few years later Acheson and Barker had established a unit funded by the Medical Research Council. Through the 1970s, Acheson looked into issues such as the effect of occupational exposures to materials like asbestos, and Barker concentrated on geographical differences in the prevalence of particular diseases. He looked at conditions such as breast cancer, kidney stones[1] and stomach cancer,[2] discovering that people in some areas of the country were at a much greater risk of getting these diseases than were people in other areas. For example, women in East Anglia were at a surprisingly high risk of getting breast cancer.

At that point, he didn't look at heart disease. This was less omission – more policy. Another MRC-funded research group was looking at heart disease as part of their work, assessing the impact of lifestyle on adult health. At the time, it seemed obvious that heart disease was a product of adult lifestyles.

The third senior person in the Southampton group was Martin Gardner. Working with Acheson, he studied thousands and thousands of death certificates spanning an 11-year period. He divided the data between the 1,366 local authorities and produced detailed atlases showing whether the numbers of people in each area who died of cancer was above or below the national average for England and Wales.[3] In their detailed work, they created maps identifying the mortality rates for 33 broad classes of cancer. The thinking behind these maps was that if you could discover areas with particularly high or low incidence of individual forms of cancer, you might subsequently be able to track down the factors causing the disease. For example, the maps showed that death rates from nasal cancer were related to the proportion of the population working in the furniture and leather industries, and that mortality from pleural mesothelioma, a form of lung cancer, was high in areas where asbestos was frequently used. In both cases the association between these sets of cause and effect had been suspected, but this added more

[1] Barker D. J. P., Morris J. A. and Margetts B. M. (1988), "Diet and renal stones in 72 areas in England and Wales", *British Journal of Urology*, 62, pp. 315–318.

[2] Coggon D., Barker D. J. P., Cole R. B. and Nelson M. (1989), "Stomach cancer and food storage", *Journal of the National Cancer Institute*. 81, pp. 1178–1182.

[3] Gardner M. J., Winter P. D., Taylor C. P. and Acheson E. D. (1983), *Atlas of Cancer Mortality in England and Wales 1970–72*, Chichester: Wiley.

evidence for the link. The maps also showed that the numbers of people dying of stomach and oesophageal cancer were lower than the national average in south east of England, but significantly higher than average in North Wales and the South Wales valleys.[4] Conversely, lung cancer was a particular problem in and around London, but much less of an issue in Wales, though for men there was a significant hot spot in the industrial zone around Newport, South Wales.[5]

Then, in 1984, Prime Minister Margaret Thatcher invited Acheson to take the post of Chief Medical Officer. "That left me holding the baby," said Barker.

Soon after, Gardner and Barker produced maps of diseases other than cancer.[6] These included maps showing the incidence of heart disease around the country – and at this point all the threads of Barker's career started to tie together. Heart disease was becoming a big issue because as populations became more wealthy the rates of heart disease increased. Their maps soon showed, however, that once again some areas had a much greater problem than did others. Barker deduced that if he could identify differences in the populations in high- and low-risk areas, he might be able to give sound advice about how to tackle this scourge.

The key idea floating around at the time was that it must be something to do with differences in the adults' lifestyles. This seemed obvious – after all, heart disease predominately affected adults. But already this was being questioned; though Barker believes there was a basic reluctance within the medical profession to really admit it. "They were falling back on ideas that heart disease must be due to many different factors. A UK government report[7] concluded that the social and geographical differences

[4] Gardner et al., Atlas, pp. 12–15.

[5] Gardner et al., Atlas, pp. 22–23.

[6] Gardner M. J., Winter P. D. and Barker D. J. P. (1984) Atlas of Mortality from Selected Diseases in England and Wales 1968–78, Chichester: Wiley.

[7] Townsend P. and Davidson N. (1982), Inequalities of Health: The Black Report. Harmondsworth: Penguin.

reflected the diffuse effects of the 'class system'," says Barker, who found this unsatisfactory in the extreme. For him invoking these concepts was as reprehensible as going back to ancient and discredited theories such as the myth that malaria was due to breathing bad air in swamps, instead of realising that it was caused by a scientifically identifiable parasite.

The heart disease map of England and Wales clearly showed that people living in the northern industrial towns and other poorer areas were at greater risk than those in more prosperous towns. But there was one startling paradox. London had plenty of poor people, but low rates of heart disease.[8]

For Barker, this indicated that wealth and poverty, on their own, were not enough to explain the disparity. Although the incidence had increased as the country became more wealthy, it was the less well-off members who were suffering most. It didn't make sense that the people in the poorer northern industrial towns that had grown up in the industrial revolution had got too much prosperity factor. Instead, Barker postulated another set of factors that were linked with poverty. He suggested that poverty makes people vulnerable to prosperity. This implies that getting prosperous is a more dangerous thing for people who are born poor. "We also suggested that this went back in time – that your vulnerability was something to do with your development," explained Barker.

On the trail of a killer

Finding a pattern in the distribution of a particular spectrum of diseases is one thing. Coming up with a theory as to why this spread of disease has occurred is also relatively easy, relative that is to finding data that can either support or falsify the hypothesis. To test their ideas, Barker and his team returned to their maps. They had all the records of how people died, but now they wanted to see what the rates of infant mortality had been at the time when these people had been born. This would give a strong

---

[8] Gardner, Winter and Barker *Atlas of Mortality* pp. 8–19.

indication of the overall health of the mothers and their infants. "There are two broad reasons why babies die. One is that they are born too weak to survive once their mother's protection is removed and they are basically small, puny and not viable. The second reason is that they are born into an unkind world with poor housing, poor hygiene, exposure to infection and over-crowding," says Barker. The first group will tend to die within a few days or weeks of birth; the second group is at risk of dying later in their first year of life.

The data that they had used to create the maps came from people who had died of heart disease aged between 55 and 74 years old. What they needed now was information that would give a clue to the living conditions into which each person had been born. Consequently, Barker and his team got hold of the infant mortality data for England and Wales for the years around which these people were born. They published their results in the medical journal the *Lancet* in 1986.[9]

This paper came to a radical conclusion that wherever you go in the country, if you know the historic data showing the proportion of infants who died around the time a person was born, you can predict their risk of heart disease as an adult. And it's a very strong prediction. Furthermore, you get the best prediction from knowing about newborn babies' deaths rather than from babies who die after the first month. Barker's earlier work with embryology told him that if you die at or around birth, it means that something in your life in the womb failed to prepare you for life outside. The deaths of very young babies were also normally linked to low birth weight.

What he had found was that people born in communities where life in the womb was tough were at greater risk of suffering heart disease later in life than were those who had a more prosperous first nine months of prenatal life. Maybe this was where heart disease begins.

[9] Barker D. J. P. and Osmond C. (1986), "Infant mortality, childhood nutrition, and ischaemic heart disease in England and Wales", *Lancet*, i, pp. 1077 – 10.

Anomalies in Lancashire

Having become interested in infant mortality in the time spanning the 1920s and 1940s, Barker started trawling the British Medical Association library, searching for anything known about the causes of infant death. He felt that this would point him towards the underlying causes of heart disease.

During this time he came across a report about three towns in Lancashire – Burnley, Nelson and Colne. All three lay close together on the western slopes of the Pennines, the hilly spine running north-south along the centre of England. In around 1914 there had been an enquiry into why mortality in Nelson was so much lower than in Burnley or Colne.[10] In 1911–13 there were 177 deaths per 1,000 births in Burnley, 130 per 1,000 in Colne, but only 87 per 1,000 in Nelson. Nelson's score was right in the middle of the national average. These differences were surprising as all three were dominated by the same industry – cotton weaving – and Nelson was sandwiched between the other two. For the six miles from the centre of Burnley through Nelson to Colne there was hardly a break in the line of houses.

When Barker compared these values of infant mortality to the death rates shown on his recently generated mortality maps he found that they matched. Burnley and Colne had much higher incidences of heart disease than Nelson. Finding the correlation was one thing – explaining it was another.

"The close proximity of the towns precluded explaining the large differences in mortality in terms of environmental variables such as rainfall. Nor is it likely that there were important differences in medical care. The hospital services for the three towns were centred on Burnley. Rather the effect of socioeconomic factors is suggested," said Barker.[11] His research pointed out that

[10] Local Government Board (1910), Thirty-ninth Annual Report, 1909–10: Supplement to the Report of the Board's Medical Officer, London: HMSO. [reported in Barker and Osmond, 1987]

[11] Barker D. J. P and Osmond C. (1987), "Inequalities in health in Britain: specific explanations in three Lancashire towns", British Medical Journal, 294 pp. 749–752.

the towns were recorded as being among the poorest in England and Wales, but that there were no differences between them. Looking at what they ate didn't offer much help either. Barker and colleagues studied the diets of 2,340 middle-aged men in the three towns, looking for telltale signs that they were eating too much fat. Instead they discovered that the men in the town with the highest rates of heart disease had the lowest intake of fat.[12] Any thought that this was a simple question of poverty was dashed by the finding that although Nelson had the best health records, it also had the greatest excess of manual workers. By any standard assessment, having high proportions of labourers should have made Nelson the most high-risk community to belong to as it would tend to have greater levels of poverty and the inhabitants would be exposed to more dangerous working conditions.

After much head-scratching, Barker headed off along a different path. He noted that of the three towns, Nelson was the youngest. While Burnley and Colne had been established in the 1700s, Nelson had only been created in the late 1800s. This had two consequences. First, the housing was much better. But second, the families living in the newer town of Nelson had been there only for one generation. Those living in Burnley and Colne were second- or third-generation industrial workers. The women living in Nelson in the early 1900s were described as "sturdier and healthier" than women in the other two towns. Barker interpreted this as evidence that poor health and physique among women before and during their pregnancy was impairing the growth and development of their babies in the womb. This poor start to life would also have continued during breast-feeding as the less healthy women would have been less capable of producing sufficient amounts of highly nutritious milk. He made a bold conclusion: this restriction in babies' growth placed them at increased risk of heart disease and stroke later in life.

[12] Cade J. E., Barker D. J. P., Margetts B. M. and Morris J. A. (1988), "Diet and inequalities in health in three English towns", *British Medical Journal*, 296 pp. 1359–1362.

Barker believed that the effect of poor housing and atrocious living conditions had gradually built up over the generations. Less-well women have less-well daughters. These daughters subsequently are in a poor shape when they come to bear children, and so starts a downward spiral. On its own the work didn't yet prove anything, but Barker was convinced that it did give insight into the kind of things that might underpin coronary heart disease. "I don't think that anybody took a whole lot of notice at the time," commented Barker with a wry smile.

In the 1800s and early 1900s the problem was that girls grew up in very bad conditions and went to work part-time at the mill when they were ten. By the time they had their own babies they had poor nutritional stores. Barker is anxious that the twenty-first-century fashion among young women to keep their bodies almost impossibly thin may have the same consequences for any children they bear.

Hertfordshire data

Up until this time, explained Barker, the work had been at the level of gross "ecological" studies. By this he meant that he had been looking at trends and behaviours of whole populations and trying to link his findings with aspects of their living conditions. He had one batch of information that gave clues about the sorts of people who were being born 50 or so years ago, and was comparing this with another batch of data collected from the sorts of people who were dying some 50 years later.

What he needed to do was to identify individuals and trace their development. That is obviously a much more demanding task because it would mean finding a group of middle-aged people who had records of their birth-weights and could be traced individually even though five or more decades had passed. If this was possible, his team could give each of these people a full health check and see how these data correlated with their individual birth-weights. The key problem was that nobody knew if there were any large batches of accurate and detailed birth records.

In 1985, while still working on the Lancashire mill town data, Barker got his daughter Mary and medical student Fiona Imrie to start searching the records offices in England and Wales, looking for any signs of birth records from the 1920s, 1930s and 1940s. He was on holiday in Cornwall when the two arrived saying, "Well, we have definitely got something in Plymouth." They had uncovered the birth records of 400 people. "It was extremely exciting," said Barker. In the event the archive wouldn't let them use the records for their research, but for Barker it proved that such records existed. There must be more elsewhere.

Next, Mary discovered that a lot of health records changed hands in 1974 when a massive programme of local government reorganisation abolished the post of medical officer of health in each town hall. The records that each of these officers had built up over the years must have been put somewhere. "I wrote to everybody who was a medical officer of health in 1974 and a woman called Dr Clark wrote back saying that she had put a lot of stuff in the Hertfordshire archives." This was a surprise, as Barker's team had already been through the Hertfordshire archives and found nothing. But knowing that the material was there they tried again. It turned out that the records hadn't been catalogued, but this time the archivist kindly found them. "It was a tremendously important find," said Barker, the excitement still resonating in his voice.

What they discovered were the records of 40,000 people born from 1911 onwards. In 1911, Hertfordshire had set up a system where a midwife attended every delivery and weighed the baby. Then a health visitor visited the mother at intervals throughout the first year and weighed the baby again at one year old. Here were all the facts and figures.

The records were ideal. The problem was going to be tracking down all these babies now that they had grown up. Then came another piece of good fortune. War can cause huge disruption, but in a strange way it can also create order. This is partly because many anxieties of infringement of personal liberties are swept away as authorities want to discover exactly who is in the

Arthur White's physique throws down the gauntlet to any who would take him on.
Reproduced with permission of Arthur White.

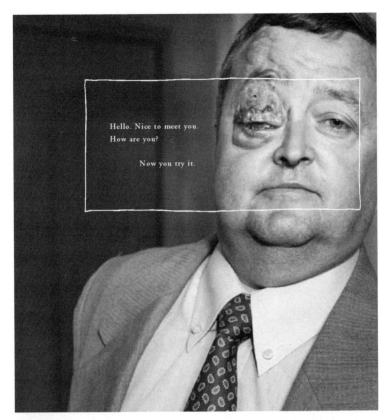

Hello. Nice to meet you.
How are you?

Now you try it.

Now, I'm the kind of guy who likes meeting new people. This is bad news for you, because I'll probably just walk right up to you and say hello. Which is the very thing you've been dreading from the moment you clapped eyes on me.

But don't worry, I'll try to put you at ease. I'm an expert at it. Let's face it, I have to be. Frankly, most people are intimidated by the way I look.

But it's quite easy. All you do is look at me for who I am instead of how I look. Ok, maybe not so easy. The best thing to do is establish eye contact, smile and maybe shake hands. If you're the shy, retiring type, I'll start the conversation.

After a couple of minutes chatting, I guarantee you will be thinking less about my features and more about what I have to say. You'll have stopped looking round the room for someone you half know, who can rescue you. Your palms will have stopped sweating and you'll feel strangely jubilant that you have overcome an irrational fear.

Before you know it, we'll be saying our goodbyes. I'll say 'Nice to meet you,' and this time, you might just agree.

David Bird was pleased to take part in a poster campaign organised by Changing Faces, hoping that it would help people get used to seeing different faces around town.

Reproduced with permission of Changing Faces. Photograph by Andy Flack. Copyright © Changing Faces.

David Bird and his wife Joanne have a mutual respect for people who are 'different', and a deep love for each other.

Eileen and Alan Piddock have been through bad times, but feel they have grown in the process.

Des and Anna Putt share a lot of humour as they live with Anna's physical disabilities. Reproduced with permission of David Chamberlain. Copyright © David Chamberlain.

While Ann Jeremiah and Judy Tabbott share their genetic heritage they have lived very different lives. They share many characteristics and mannerisms but also have clear differences of character. Genetics is clearly important, but is only part of their story. Reproduced with permission of David Jeremiah.

David Barker pulls a book from the set of data held in the Hertfordshire archives in 1986 – data that would change his life, and make us realise that heart disease starts in our history.
Reproduced with permission of David Mansell, photographer.

This photograph is an important part of Christine Whipp's story, telling her where she started in life – in the arms of Nurse Puncheon.

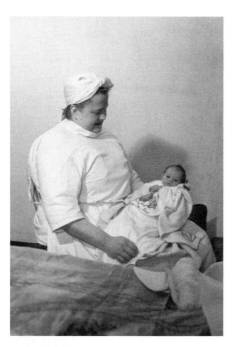

Her parents' wedding held more significance for Christine before she discovered that Wilf was not her father.

James Lovelock's Gaia theory will either be forgotten in a few years' time, or rank alongside evolution as one of the major stepping-stones in our understanding of life.

country, and keep tabs on where they are living. On the outbreak of war in 1939 a Britain-wide census was carried out. It covered the entire nation with the purpose of identifying foreigners. Everyone had to register, be they tramps who slept on the beach, or landowners in castles. There was an incentive – you couldn't get a ration book without being on the list. After the war this list became the basis of the National Health Service Register. Barker's team set about matching birth records with census listings, and using this as a stepping-stone to getting in touch with the individuals for whom they had birth records.

The whole exercise was a massive gamble. There was no way of knowing whether they would succeed in tracing enough individuals, and even then there was no guarantee that the original data were worth using. After all, the nurses were only weighing babies in cottages by candlelight on crummy old scales. Thankfully, for Barker, most of the people who had been born in Hertfordshire in the 1920s hadn't moved far. Tracing them was a huge exercise involving trawling through baptismal and marriage records, but the day came in 1989 when Barker was convinced he had traced enough. Some were dead, but that was not a disaster for the research, because in those cases they still had records showing the cause of death.

"So statistician Clive Osmond and I effectively pushed the button one day," Barker explained. There in front of them were the results of complex calculations based on all the data they had collected. They were thrilled. The relationship between heart disease and birth weight was clearly there. But this time, the relationship with weight at one year was huge. "Awe-inspiring actually," said Barker. "It was absolutely amazing."

At the time many people were quick to criticise his work, suggesting that the whole of the results were simply the result of people making inaccurate measurements. Barker's quip was that if measurement error can give such massive predictions of the likelihood of an individual suffering from coronary heart disease, then let's have some more bad measurements! Obviously, he was convinced that the measurements were sound and that he was on the trail of something big. A second academic report was published in the *Lancet*.[13] That really set the agenda.

To some in the medical community, Barker's findings were uncomfortable. His ideas were calling into question much of the standard advice given about lifestyles. For years people had been concentrating on a person's adult life as a way of staving off diseases caused predominately from living in a prosperous society. Now Barker was saying that the critical period was the first year of life and that, more than this, your growth in the womb was also vitally important. If you had difficulty growing in the womb, then you would be prone to heart disease. As time passed, data came from other studies in Sweden, then Finland and America, all of which looked at the issue from different angles, but came to the same conclusion. Growth comes shining through as the critical factor, no matter what the social circumstances. But then, Barker realised that in the real world this is linked back to a person's social circumstances, in that poor living conditions will impede growth.[14]

Placentas in Preston

To help make sense of the Hertfordshire records, Barker had employed Jackie Ariouat, an Oxford historian. She had originally come from the northern industrial town of Preston and she consequently started looking to see if there were similar sets of data in her hometown. She wasn't disappointed. One magnificent building in Preston had housed the Sharroe Green workhouse and its hospital. The hospital had a maternity unit and the midwives had diligently collected records not only of each baby's

---

[13] Barker D. J. P., Winter P. D., Osmond C., Margetts B. and Simmons S. J. (1989), "Weight in infancy and death from ischaemic heart disease", *Lancet* ii, pp. 577–580.

[14] At the time of writing this book in 2003, the Hertfordshire data set included 4,000 people, all of whom had been monitored and had donated DNA samples. The stream of information from this database has only just begun.

weight, but also each infant's length and head circumference. On top of this they had weighed each placenta.

As they were analysing the Hertfordshire data they were aware that birth weight alone was not going to answer all the questions, and Barker wondered if this additional information might help. So he employed a nurse to go round and just measure the blood pressures of all the people they could find who had been born in this hospital. "When she started off I said, 'When you have done 250, we'll have a look at the results.' "

Once again the results showed that low birth weight was related to raised blood pressure. But what leapt out of the data was the fact that a combination of low birth weight and high placental weight gave an incredibly strong prediction of raised blood pressure as an adult. The placenta is the organ that transmits gasses and nutrients between the mother's and the fetus' bloodstream, but there was no obvious reason why its size should be important. "The results were huge, very, very impressive," claimed Barker, brimming over with pride in this achievement.

They decided to double the size of the survey, and to make things quicker, Barker joined in data-collecting. "It was just before Christmas and it was fantastically good fun. We were knocking on the doors of a bunch of people who had last been together when they were newborn babies in Sharroe Green Hospital in the late 1930s. Some of them were living in extreme poverty; some of them had become very rich. It was huge, huge fun," chuckled Barker. "We stayed in a hotel just opposite Sharroe Green, £10 a night. The owner used to turn the heating off during the day, but we were in during the day and out in the evenings, because most of the people we wanted to see were at work. My room had this great big bed and we all used to pile in because it was so bloody cold!" They landed up with complete records for 449 men and women aged 46 to 54. All still lived in Lancashire. The findings confirmed the startling evidence. Life in the womb had an important effect on blood pressure in adults.

I reminded Barker that it was at this stage that I had met him for the first time. At that point I was working at University College London, and Barker called in one particular lunchtime to show us his results. I remember sitting there in disbelief. Surely no one would believe the idea that being born small with a large placenta would start you on the track to middle-age heart disease. We had viewed the data with a mixture of scepticism and disbelief. We, after all, were physiologists and were used to performing carefully controlled experiments, rather than wandering the streets collecting numbers by knocking on doors. At the same time we had to acknowledge that the data were amazing, although we preferred to put the idea out of our minds and get on with our "serious science". But Barker was convinced, and wasn't put off by our thinly veiled doubt. When he left we all had a good laugh.

Our research team wasn't the only one to think that Barker was beginning to lose the plot. He soon found that the *Lancet* wouldn't publish this new set of data. "So I rang the editor of the *British Medical Journal*. He asked, 'Is this new?' I said, this is amazingly new, you are predicting hypertension from placental weight and birth weight and these are huge predictions." The *British Medical Journal* published it.[15]

Publishing your work does a number of things. It enables you to stake your claim to a piece of work and gives you the academic credibility that accompanies this territorial demarcation. It also lets other people see what you are doing. A few weeks later Barker received a letter from Australian researcher Geoffrey Robinson. This gentle giant of a man invited him to travel down under so that he could explain what he felt was the link. Robinson was an obstetrician, but he also was very used to doing research work with sheep. When Barker arrived, Robinson explained that if a lamb fetus is undernourished it will, in certain condition, enlarge the placenta. This gives it the ability to extract more food

---

[15] Barker D. J. P., Bull A. R., Osmond C. and Simmonds S. J. (1990), "Fetal and placental size and risk of hypertension in adult life", *British Medical Journal*, 303 pp. 671–675.

from the mother – it is like making a bigger gate to let in more nutrients. But the story is not simple, because this only happens if the mother is well nourished through the pregnancy. Once again, the link became clear – children born to mothers who had been poorly nourished before becoming pregnant, but reasonably well nourished through the pregnancy, were at increased risk of heart disease.

The theory that started developing here is that a mother who is undernourished before pregnancy will in some way prepare to nourish any developing fetus in a nutritionally impoverished environment. If the mother suddenly eats well during the pregnancy, this pre-planning may mean that the fetus is potentially oversupplied with particular nutrients. Alternatively it could be that the fetus is presented with inappropriate levels of hormones that aim to enable the fetus to scavenge all it can. The data showed that something was occurring, and the understanding of what exactly this was would have to catch up later.

In the meantime Barker's team had discovered records in Sheffield that were even more marvellous than Preston's data. The Sheffield records also included measurements of the circumference of the chest and the circumference of the abdomen where the liver lies. A baby in the womb uses their liver as the biological equivalent of a larder – it is a store that is packed with energy, laid down ready to give the baby reserves just in case life is tough when they pop out. The liver circumference turned out to be a wonderful predictor of future blood cholesterol levels and an amazingly powerful predictor of death from coronary heart disease. A baby with a small liver at birth had high cholesterol later in life. "A simple measurement round the tummy of a newborn baby was predicting cholesterol levels 50 years later – incredible!" said Barker.

To India and beyond

If this historic element of our existence is a general principle of the ways that genes and environment interact as we develop then,

Barker surmised, he should find the same sorts of things occurring in all parts of the world. Contacts and circumstances once more played into his hand.

"I was asked to go to Belgium in 1990 to teach young epidemiologists about cardiovascular epidemiology. We used to fool around in the bar in the evening and I made friends with Srinath Reddy, a young Indian cardiologist. I taught him to play darts and we used to beat the Germans and drive them to despair!" said Barker with thinly disguised pleasure.

Reddy became interested in ideas about the early origins of coronary heart disease. For Reddy, the idea made basic sense. It fitted well with Indian cultural ideas and beliefs, which emphasise that your life goes back into the dark recesses of history. According to Indian culture, what happens to you now is ordained long ago. He invited Barker to a meeting in Delhi. Barker took with him Caroline Fall, a paediatrician whose father had been posted to India as a British officer, and commanded an Indian regiment during World War II. Consequently, Fall had many good friends and "godfathers" in India and loved the country.

Arriving in India, Barker found himself addressing a group of reporters. Heart disease is becoming increasingly common in India and there is considerable public interest and concern. Barker explained that in the West mainstream medical opinion held that heart disease is something to do with the way you live your life as an adult. The reporters just couldn't believe it. This seemed to be a ridiculous concept – only a very naive civilisation would take on ideas like that. They just imagined that once again the Americans were trying to impose some form of control over them, by telling them how to live their lives. To these reporters, Barker's idea of tying heart disease back to life in the womb simply seemed obvious. "Really what they were saying is that you should go back. It is not about the baby, it's about the baby's mother and father – it has got to go back more than one generation," commented Barker.

Soon after the meeting, Barker got Fall set up in India with her own funding. "She has now got a Wellcome Trust programme grant and the Indian studies have become very important. They have confirmed the Hertfordshire findings." This started to confirm the universal nature of Barker's hypothesis and Fall's programme of research now includes work in Vellore, Mysore, Mumbai, Pune and Delhi.

Barker recalls that the early days of the Indian operation were hard. They were in the familiar situation of needing to find accurate birth statistics that could be linked to currently identifiable adults. It was a bit of a tall order in India. Fall wrote to hundreds of hospitals and found that the Holdsworth Memorial Hospital in Mysore had wonderful records of the kind that they had in Preston. The hospital had been built as a mission hospital for women and babies and been deliberately placed in the middle of a very poor area. Once again they were faced with the standard problem. "How do you trace the adults in India, in a slum – how the hell do you find the people?" asked Barker rhetorically. The answer is that you do a census of the entire slum. German doctor Claudia Stein and a team of eager researchers went from house to house asking whether the residents had been born in the hospital and doing a health check on any they found.

In about 1997 the results were through – again confirming Barker's key hypothesis. Low birth weight was linked to heart disease.

Barker's next break came through links that he had been building in China. He had been going to China since 1986 when Shirley Williams MP set up a cultural exchange between the two countries. Over the years he had asked about records, but always been given a short answer. No – none existed. But then the Queen went to China in the early 1990s, taking with her one of the surgeons for the royal family, Norman Blacklock (now Sir Norman). He made friends with Wu Jie Ping, the President of the Chinese Academy of Medicine, partly because they shared a common interest in surgery associated with the urinary tract.

A few years later Wu Jie Ping visited England, and was introduced to Barker and his data. Wu Jie Ping turned out to be a very powerful person – he had been the doctor to Chiang K'ai-shek, the leader of the Chinese National People's Party who had fled into exile in Taiwan in 1949. Wu Jie Ping offered to find Barker the data he needed. "In India we had sorted everything from the bottom upwards, but in China we sorted it from the top down," explained Barker.

He soon received a message that there were one million records in the attic of the Peking Union Memorial Hospital and, what's more, the records were in English. Chinese doctors had done the same meticulous measuring of babies in this hospital as their English counterparts in Preston. Wu Jie Ping had known about their existence because he had trained there when the hospital was owned by the Rockefeller Foundation. It was the place where the very best and brightest Chinese medical students went to study. These were the days before Mao Tse-tung's Chinese People's Republic took control of the hospital.

Wu Jie Ping remembered that when a baby was born, doctors not only took a host of physical measurements, but the placenta had to be drawn by a medical student, both from the top and from the side. This was compulsory for every baby. All sorts of things were recorded, including fingerprints and footprints. As Barker said,

> This was an amazing, amazing discovery and, in an organised nation like China, there seemed little problem in tracing the people associated with each record. Indeed, within a week we found a whole bunch of people who had been born in that hospital, despite the famines and the cultural revolution and the banishment to the countryside of the intelligentsia, all of which had driven whole populations into the hills outside Beijing. It was amazing.

Once again the data linked low birth weight with high risk of heart disease. It became increasingly clear – this was a universal phenomenon.

Helsinki takes us back further

Data from the different parts of the world were now clearly showing that a life of nutritional poverty in the womb affects your adult life. But heart disease is a disease of Western prosperity. Barker still had some work to do to show how the two were linked. He was increasingly convinced that the nub of the problem came when fetuses and very young babies grew in poverty but were exposed to prosperity in later life – these people were the ones at greatest risk of so-called Western disease. This still left open an issue. What was the prosperity factor?

Barker began to dream of locating a population of people who had experienced a poor nutritional environment in early life, but had then been introduced to a "prosperous" diet. He was well aware that babies have an incredible ability to perform catch-up growth that could easily mask a malnourished start in life. If a child gets ill or is malnourished for, say, six months, its growth will falter. But if it is subsequently nourished well it can grow at an accelerated rate, faster than it normally would. Could it be that this catch-up growth is part of the problem?

But where was Barker going to find a group of people starved in the womb but well-fed thereafter? The answer was Helsinki.

At a meeting in Brussels in 1995 Barker sat next to a man called Johannes Eriksson. He is a diabetes specialist working in Helsinki. After a brief conversation the two parted company, but then a few weeks later Barker received a letter, this time inviting him to cast his eyes over a set of data in Finland. From around 1925 onwards, all the children in school in Helsinki had their heights and weights measured every six months by two school doctors and that all that data had been preserved. In addition, from 1934 the Finnish government set up numerous child welfare clinics and preserved all the data from these health centres in the city archives. On top of this, the University Hospital of Helsinki had birth records going back to 1900, all accompanied with detailed measurements.

A month later Barker was on his way to Finland. "The data is astonishing – it is the best-documented group of people in the world." The archives record each person's size at birth. This can be linked to each person's school records, which have details about their home, including information about how many people lived in it. Because everyone has an identification number this can be linked to the registers that record medication each person has been given during their life.

Baker chose to study a group of men born between 1924 and 1933, and the comprehensive nature of the data enabled his team to reach a clear conclusion. A person's risk of dying from heart disease was greatly increased if they had poor nutrition before birth, and then grew rapidly in the first seven years of life. Barker was refining his theory. It appeared that your body takes note of its nutritional environment while it is developing in the womb. If it detects that life is harsh, it plans ahead, and sets itself up so that it is best prepared for life on a poor diet. But this means that it is ill-prepared for a rich diet, and this will in fact be harmful. Conversely an infant that is well-nourished in the womb will prepare for a life of plenty. This person will then be in a better position to enjoy the supposed classic rich Western hamburger diet, secure in the knowledge their body is built for it.

So, when do critical events in your life occur that shape who you are? The answer has now been stretched back not just to the first weeks after birth, but to the first few days and weeks after sperm and egg meet at fertilisation.

By this time Barker had teamed up with a small group of scientists who specialise in studying how fetuses grow in the womb. Their message was simple. If you want to find out about the influences on the baby in the womb, it all starts early, very early. The tendency for doctors to think they can almost ignore early pregnancy because the embryo is so small is foolish. Yes, in late pregnancy a growing baby places heavy demands on the mother, but an early embryo is profoundly sensitive to the hormonal and metabolic environment within the mother, and this will be influenced by her physical condition and diet.

This is now Barker's area of study. But the thinking has led to one more startling discovery. All of a female embryo's eggs are in place inside her tiny ovary after only 14 weeks of growth. These immature eggs are therefore exposed to the conditions inside the womb. Later in life, when the baby has matured and is having babies of her own, she uses those eggs. Emerging from the Helsinki data are results showing that a person's risk of heart disease is not only affected by the environment they experienced in the womb, but also by the environment that the person's mother experienced while she was in her mother's womb. According to Barker, this intergenerational succession is known in animal husbandry. "If you do something to a female animal you affect at least two generations," he pointed out.

So now we are faced with an even more extreme concept – that the physical condition your grandmother was in and the diet she ate while your mother was in the womb has a strong influence over your likelihood of suffering from heart disease at age 50. When did life-shaping events take place? An entire generation before you were a twinkle in your father's eye.

This now leads to a number of questions – some of them have been grounded with answers, others remain airborne. Can the human embryo alter its development in a manner that has been seen in other mammalian embryos? The answer to this one is, yes, it can. But how about the egg? Fertility specialists around the world who create human embryos in Petri dishes are becoming increasingly aware that the egg responds to its environment before it is fertilised. There is now no reason to assume that the egg is in some state of unresponsive, suspended animation while it is in the ovary, and only becomes responsive once it is fertilised by a sperm. In fact, it has got to the point that fertility specialists are now questioning whether the way they treat eggs prior to fertilisation and during the first stages of embryonic development may have an impact on a wide range of diseases that includes diabetes, heart disease, stroke, depression and schizophrenia.[16] Fertility specialists now even think that they may get valuable

---

[16] Schatten G. P. "Safeguarding ART", *Nature Cell Biology and Nature Medicine*, www.nature.com/fertility s19-s22.

data by carefully monitoring the health of children born as a result of fertility treatments. As some of the oldest children conceived this way have now given birth to children of their own, there is also the possibility of seeing whether any of the influences that come from starting out in a Petri dish are passed on to future generations.

Barker is passionately concerned that so much attention within research is devoted to genetics, in the hope that studying our genes will enable us to fix everything. His view is that knowing the spectrum of mutations in the genetic pack of cards dealt out to an individual at fertilisation is of little value, unless, that is, you have a detailed knowledge of the person's history. It's not the genes that are important, as much as the way that the genes act in response to the historic environment they work within.

So when did you begin? What were the defining moments that established your health characteristics? Was it your childhood diet, or what you ate last week? Probably neither. There is now compelling evidence that you, your personal history, started at least two generations before you were born. You are a historic being.

CHAPTER FIVE

a related being

Christine Whipp lives in a pleasant redbrick semi-detached home with a living-room, kitchen and cloakroom downstairs and three bedrooms and a bathroom upstairs. The house is part of a small estate on the edge of the medium-sized market town of Honiton in the south west of England. She is of average height, has blue eyes and her straight fair hair almost reaches her shoulders. Her two grown-up daughters have left home, so she now shares the house with her husband and a gang of cats, young and old. The walls and mantelpieces have photos of family, and mementoes of happy days. On the surface there is every appearance of normality.

While Christine is every bit a normal human being, her story is not, and goes some way to revealing the extent to which our perception of ourselves is influenced by, even based on, our family. In Christine's case, she didn't discover the true nature of her family identity until she was 41, but she had long suspected that all was not quite what she had been led to believe.

"I made a very bad start in life," laughed Christine as we sat down to start recording her life story. "I was born on 1 April 1955. My mother wanted me to be born on 31 March, but I hung about and was born five minutes into April. I was an April fool. She found that most embarrassing." Christine leaned forward to show me a fading black-and-white photograph of a newborn baby, wrapped in white and held by an enormous nurse, Nurse Puncheon. "What an amazingly fat lady, she was absolutely huge," said Christine with no hint of criticism. But for Christine's sense of identity, this has become a pivotal photo. It was taken in the front bedroom of her parent's home in Bridport, five days after she was born. "There," she said.

> It's been quite important that I was born there, visibly born there, because it told me all my life that I hadn't been adopted. If people have known you were born, and there is a story about how you were born, then you can't have been adopted. This is me with Nurse Puncheon, the nurse who went out delivering babies for local people. My mother had

a normal pregnancy with a big bump and I was a normal ordinary person. I was her child.

Nurse Puncheon was a link to normality. "I knew her. It was a small market town. I would bump into Nurse Puncheon and she would say, 'Hello Christine. I remember when I delivered you.' There was no question that I was adopted. I belonged to those people in that bungalow."

Christine's father, Wilfred Bartlett, was not a well man. He had had a very bad case of mumps when he was 13 and was an insulin-dependent diabetic. This is serious enough in the twenty-first century, but in the 1950s medical insulin was less pure and it was harder to control sugar levels. If you let blood sugar rise too high and too often, the small blood vessels in your body suffer. You damage organs all the way from your eyes to your toes. Over time the accumulated disruption can even kill you. He was about 15 years older than was Christine's mother and would die when Christine was only six.

As Christine grew up, she learnt about her heritage within the Bridport area. Her mother's family were respectable and all lived in the town – her mother's father had been a policeman during the war. He had worked as a master-plasterer and she could point to many houses that he had been part of. Her father's family also lived in the town and had roots in the village of Shipton Gorge near Bridport that went back for four or five hundred years. They were quite well-to-do people with their own construction company – building homes, furniture and coffins. "My parents had wanted a son. My father was the only male in his generation and they needed a son to keep the name and run the family business. I was a girl," she commented sourly.

Christine's parents were Mr and Mrs Average. They got married and the splendid black-and-white photo records that it was a big fancy wedding even though it was 1948 and there was still rationing. Wilfred was pristine with his black hair immaculately parted. Barbara looked gorgeous in a long white dress. "We were on the up – we even had our own mortgage, which was no easy thing given that he was self-employed and had diabetes," Christine said.

"People knew that mother had had difficulty getting pregnant, because she had been going to see a posh gynaecologist lady in Exeter. She'd even had a minor operation at Portway Hospital. Obviously the problem was solved and the next thing was a pregnancy," recounted Christine, handing me her original birth certificate. "It says here, anyone falsifying any of the particulars on this certificate, or using a certificate knowing it to be false, is liable to prosecution." And indeed there was nothing false on the birth certificate, because it was the short-form certificate that simply confirmed her name and date of birth. "I am the person named on there – but I am not the person described on the full certificate, the certificate that they didn't buy," she added with a conspiratorial tone in her voice.

> When you are a child you take a lot for granted, and you absorb your identity from your parents. There is this woman saying, "Say Mum-mum" and a man saying, "Say Dad-dad". You are presented with your aunts and uncles. There are friends and grannies all around you. You are building who you are through the people you are interacting with. But I knew that my mother didn't really like me. You know something is wrong when your mother comes at you with a wooden spoon to attack. You know something is wrong.

As a baby, Christine cried a lot of the time. "I cried so much I spent my first summer in the pram at the bottom of the garden – as far away as possible from the house," said Christine handing me another photo.

> As you can see, it was a very hot summer and she would put a canopy up, but as the sun went round, my feet left the shade and became practically burnt. Apparently a friend came down from London for the summer, looked in the pram and said in a Cockney accent, "Ow my Gawd, she's a

littl' nigger!'' It was meant as a joke, but can you imagine
how my mother must have felt – she must have thought,
"Oh my God, who was the father?"

Christine said that her skin has always been dark, at least, much
darker than her parents'. It's very likely that they all quickly
laughed off the comment and that her mother made a better job
of keeping Christine's feet in the shade.

Obviously Christine was too young to remember that story for
herself – it's one that she was told as she grew up – but as life
went on there were other events that she can remember, first-
hand.

When I was about five or six, I remember the afternoon
clearly, it was just after Christmas and I went off to the
Sunday school Christmas party. I was all dressed up in a
new blue dress that my grandparents had bought from the
Freeman's catalogue and a little white jacket that a friend
had knitted from angora wool – very trendy at the time. I
had a giant pink petticoat underneath that made the whole
dress balloon out when I stood up. Uncle Norman, my
father's brother-in-law, came and took my photograph
before I went.

You know how these church dos are; organised by little
old ladies busying around. Well, a couple of these old ladies
were clearly interested in me, and asked, "why don't you
come in here a moment and help us get a couple of chairs?"
It seemed a reasonable enough request, so I went. Once in
the room, however, one of them stood me on a low table
and walked around commenting on my nice dress. The
other suddenly said, "Well, who do she look like, then?"
The other replied, "Oh, I don't know – she do look like her
mother." But they seemed unhappy about that conclusion –
but then who the hell did they want me to look like? They
had worked it out. Poor old Wilf, he wasn't really up to it –
what with the mumps and the diabetes. So where did Barb
get this child from? Whose was she?

From this point Christine claims that her mind set off on a secret journey. A journey largely based in fantasy, but one that was fuelled by a conviction that she was not the person she had been told she was. At school a few years later, aged about eight, the teacher was showing the class a big old globe. She was talking about the seasons and how the Earth is tilted on its axis – how the sun shines so that it is cold at the poles but warm at the equator. "Boring old stuff – I started nodding off and staring at the clouds out of the high windows in the Victorian classroom. Then one of my classmates called out, 'Miss, if we dug a hole in the Earth, would we come out at Australia?'"

Christine's teacher pointed out that, no, they would come out in the ocean near to New Zealand. She went on to say that New Zealand was an island that had many similarities to Britain and at that point Christine's mind made a fantastic leap of childhood logic. "Suddenly it clicked – I know now – the stork was flying and took the wrong turn at the equator and dropped me off here instead of in New Zealand where I should have been. It was a 'eureka' moment. I was with the wrong people in the wrong place." To an adult it may seem an absurd thought, but it shows the extent to which Christine felt that she couldn't fit in with the life story she was trying to live. Even so, the thought was so uncomfortable that she deliberately shut it out of her mind.

Christine didn't challenge anyone at that point. Even though she felt uncomfortable with herself, she didn't want to discover that she didn't belong. That would have been too much to bear. And what was she going to do? Go to her mother and accuse her of ... well, she wasn't quite sure what of. By that time Wilfred had died and her mother had remarried. Her mother was not making life easy, even though her new stepfather did his best to be generous. Christine was desperate to be wanted and loved. She was working hard to be accepted, and this was complicated by the arrival of a new half-sister. "If I had said anything, it would have been a bad time to say it."

Aged 13, Christine started "seeing" Michael Whipp, the young lad who soon became her husband, and at last she had someone to confide in. Five years later, and she was married with her first baby, Nicole; three years after that and she had her second baby girl, Justine. Michael joined in with trying to make sense of the situation. His view was that Christine must have been adopted and that, maybe, the pictures of the baby were of another child who died. Maybe Christine's parents had adopted her as a replacement.

At 25 her level of understanding was increasing. She had discovered about her father's mumps and the chance of that making a man infertile. She had discovered that if diabetes is not treated well it could also cause impotence. "By the law of averages, it was looking less and less likely that Wilf had fathered me," she concluded.

She started doing family history research – genealogy.

I've always been into history. I found out all about the Bartletts; they were nice people, rock-solid members of their village community. They had never put a foot wrong. They had been business people, farmers, salt-of-the-earth people. But I didn't feel I belonged. I wanted to belong, but it didn't work.

Being a small town that had had a static population for generations, Bridport was a place where your family name was important. Coming from a well-established family gave people a standing in the community. Some families were Baptists, others Methodists. They wouldn't have a lot to do with each other. "There is a lot to having a name, it does define you – name, rank and serial number," Christine said.

I was trying to bolster up the foundations of where I had come from. To get on paper that these are my people. I drew out family trees, put it all together. But I kind of knew it wasn't me. Then came the day that we were sat at my mum's for Sunday tea; we went there each week. There was me, Michael, Justine here, Nicole there, my half-sister where she always sat, my stepfather in his place, Ginge, and Mother sat on the other side next to the kitchen so she could run in and out. She suddenly turned to me and said in a hoarse, but loud, whisper, "There's something about your past that you don't know." No one else appeared to hear above the noise of the meal.

Christine asked what it was and got no answer, but at last she felt relief. There was a secret. There was some complexity about her identity. It must be something fundamental about who she was.

Shortly after this she decided to send off for her full birth certificate. It dawned on her and Michael that this was the piece of paper that would tell her the truth. In October 1991 it arrived. Christine sat down before opening the envelope, anticipating a moment of great revelation. Instead, she saw the information that she had always known, but it was the same information that she had always doubted. Once again she was being told that Wilfred Thomas Bartlett was her father, and that Barbara May Bartlett was her mother. The certificate had been filled in by Wilfred.

The more she stared at it, the more convinced she became that it was not true. There was something wrong. Or at least, something was not right. "But every time I questioned whether it was true I realised that I was sounding like a psychopathic lunatic – it had to be true. It even said on the bottom that you were not allowed to falsify it. But I knew it wasn't correct."

The best theory that Christine could come up with was that her mother must have been raped and that there had been a cover-up to avoid embarrassment. Her mother had always said that she had been called "Christine" for a reason, though over the years changed her story about that reason. One of the early reasons was that she had been named after a policewoman who lived in

East Street. Christine started to wonder whether this woman had handled the case.

It also fitted in with the birth certificate – if there had been a rape then everyone would have wanted to cover it up quietly to protect me. I wondered about looking up in the newspaper archives to see if there had been a rape in Bridport at that sort of time. Soon after that my stepfather died and my relationship with my mother deteriorated further.

In the end, Christine wrote a letter saying that if her mother would tell her the "secret" then she would go away and never trouble her again. Christine felt that her mother would think that this was a good trade-off.

"She wrote back. It was a normal morning. Michael hadn't got up yet; my daughter was sat at the table eating Rice Krispies. And I read it out. 'You were conceived by donor insemination.'" Her daughter was "gobsmacked" and asked what it meant. Christine only knew because a boy at her school had had a father who was an inseminator of cattle. "It always made us laugh," she chuckled.

I thought, that's it … that's it! I'm not mad at all. I always was somebody else. I'm not crazy. It made so much sense. No wonder she hated me so much. I was now in a new area of exploration. Regardless of what the letter said, I was still the same person I had been before I opened the envelope. I was Michael's wife, Justine and Nicole's mother, Daniel's grandmother. I was Mrs Whipp. But at the same time I was somebody else. All of a sudden I was ready to grow into the person I had always been, but didn't know I was. And who was Christine?

By this time I'd already been a lot of people. I'd been Christine Bartlett to start with, then at the age of ten I'd had my name changed to my stepfather's surname. That in itself was something of an identity crisis and done without any reference to me – I was just handed the deed poll on a plate. It didn't help that my mother cut my waist-length hair short at the same time. Then I married and became Christine Whipp. In a way, these all prepared me for this third time that I had a major identity change.

That was 1996. Since then Christine has tried tirelessly to discover the identity of the donor. She traced the clinic. It was the place where her mother had gone openly for gynaecological treatment early in her first marriage. Run by Dr Margaret Jackson, this private clinic claims to have initiated 483 babies through donor insemination between 1940 and 1981.

"Most people think that donated semen always comes from medical students – so the first mental picture I had of my father was a medical student with a stethoscope around his neck. But then Exeter didn't have a medical school," she informed me. In fact, Dr Jackson had obtained the sperm from friends and senior academics at the university that she knew personally. The donors were mainly married men with families of their own.

Jackson had died ten years before Christine received the revealing letter, and there are claims that all of her records were subsequently destroyed in a local incinerator. These were meticulous records, as parents were asked once a year to send a letter with a photograph. But they are all gone so it looks as if Christine is doomed never to know her biological origins. But her search goes on. She has now narrowed it down to a list of about a dozen men who knew Jackson in 1954, some of whom are still alive. So there is still a slim hope that she might meet her biological father one day, but it seems unlikely.

"It's strange to think that your mother never met your father. I've been told that not only is he anonymous, and that was set up deliberately, but even if there were records of his name, I am not entitled to know it." Christine was getting visibly angry.

It is one thing if this anonymity was accidental, but quite another to say that even if they had the information they would still deliberately withhold it. I really feel like a subspecies. Why do I have to be treated differently? Everyone else is allowed to know who their parents are. Even if you were adopted you would be allowed to know their names, but not me.

It's important to know where you have come from – I am part of a continuum, and that continuum has been severed, and that apparently is OK, but it is not OK with me. I want to know who these people are, what made them tick, what their politics were, what they were doing in the war. How can I feel proud of people I don't know? I could take pride in the Bartlett name. These were people who had built houses. My grandfather found a pulpit in an old church in Charmouth that John Wesley had preached in. He took that tatty old pulpit, did it up and put it in the Methodist church in Bridport. That pulpit is there today – that's the pulpit my grandfather worked on – or I thought it was. Now he is just some bloke. That is somebody else's history, somebody else's family. But I want to have my own family to be proud of. I want to know where I belong. My family tree starts with me. It's a stunted bush. It just goes forward; it doesn't go back. I don't feel that I am grounded in the way that I should be.

The importance of family ties goes two ways in that a parent gains from having a healthy open relationship with their offspring, and can be damaged if this aspect of life goes wrong. Christine believes that the dysfunctional relationship between her and her mother deeply affected her mother's life. Her mother would be angry if Christine "failed" in some way, and jealous of her if she succeeded. According to Christine, the resentment ate into her mother's character. When Christine's O-level exam results came through, her mother was jealous of the praise Christine received, but displayed it in scorn towards Christine, taunting her because she had got only one top grade in the tally. Her mother's anger turned to fury when Christine's stepfather bought Christine a small watch to recognise her achievement. Later he risked

Barbara's wrath by fitting a rolled-gold bracelet strap to the watch as a reward for Christine passing her three A-levels.

"Then I made it worse by having a baby at the age of 18 in the usual way with good old rumpy-pumpy. And I didn't have to go and ask someone's permission."

The years of anger and deception eventually destroyed Christine's mother, who spent her last years in a secure psycho-geriatric unit as she had become severely violent with psychopathic tendencies.

"She broke her hip trying to escape – I am sure she was on her way to kill me. She used to sit in her hospital ward, rocking back and forward, saying, 'It were a bad thing I done, I done a bad thing.' There was something going on in that head – it was all in there," said Christine.

Not alone

If Christine's story were an isolated case you could argue that the situation was so extreme it tells us little of use about what it is to be human. But she is one of many. Official records show that in the UK alone, between 1990 and 2000 around 18,000 children were born as a result of donor insemination (DI). This is likely to be an underestimate as it accounts only for births organised through registered clinics. There are plenty of rumours of friends helping friends via unofficial routes.

In a letter to the academic journal *Human Fertility*, middle-aged Londoner David Gollancz explained the impact of discovering that his beginnings were more complex and less certain than he had always been led to believe.[1] Again he found that the discovery shattered his self-image and left him questioning quite who he was.

David describes how in the spring of 1965, when he was 12 years old, his father came along one evening and suggested

[1] Gollancz D. (2001), "Donor insemination: a question of rights", *Human Fertility*, 4, pp. 164–167.

they went into his bedroom because he had something important to say. David sat on his bed, while his father stood, obviously feeling uncomfortable. His father then explained he and his wife had spent many years trying to have a baby before they discovered that the problem was that he was not producing many sperm – in effect he was infertile. The doctors they talked with had advised them to think about using artificial insemination by donor (AID).

It took a few attempts, but eventually his mother became pregnant and his sister was born. The following years saw them try repeatedly for another child, but every attempt ended in an early miscarriage. Eventually their persistence was rewarded, and David was born.

David's father went on to say that the donor used to create him was different from the one who gave the sperm that created his sister. Both obviously still started from their mother's eggs, so they were, in effect, biological half-siblings. The identity of the donors was unknown, in fact maintaining this anonymity had been a condition of being accepted for treatment. The only thing they knew was that, like David's father, both donors were Jewish and were successful men. They were also married and had their own children. David's parents had also promised never to tell their children that they had been conceived by AID. They were breaking this promise because they thought that he and his sister should know the truth.

"In one breath, as it were, I learned that my father was not my father, my sister was my half-sister, I had been conceived by means of a medical procedure, and I could never know who my progenitor was," wrote David.

People always ask me how I felt. Six years earlier I had been hit by a car as I crossed the road. It hit me in the side. I was flung in the air and landed on my back in the road, without either losing consciousness or feeling any pain (in fact I escaped with a chestful of broken ribs). Being told that I had been conceived artificially using a stranger's sperm was like being hit by a train. It didn't hurt. I wasn't angry or grief-stricken or excited. I felt annihilated. It felt as though I had been told something immensely important but meaningless: somehow, the most important thing I had ever been told was empty of content.

As happened to Christine, the historic linkage via David's genealogy had been severed; or at least half of it. His mother's line of descent remained intact. Yes, the nurturing element that had gone into shaping him was unchanged, although trust in his parents' ability to tell the truth at all times was presumably called into question. He was, however, deeply aware that nurturing was only one part of the story. David explained in his paper:

> It seems unlikely that any sensible person would now advance the theory that genetic inheritance counts for nothing in human development. Even the most die-hard advocate of nurture-over-nature cannot deny the relevance of genetics to physical health and most people would accept that an individual's personal development is the story of the interaction of genetic predisposition with environment. Most people might also accept that an essential ingredient in personal development is the process of accounting and recounting: explaining ourselves to ourselves through a continuous process of storytelling, through which we accept or repudiate responsibility for our lives, entrench or modify our values, and describe celebrate or deny ourselves and our lives. Much of the process of accounting and recounting is done by reference to family: "I am like this because my father is like that", "I can do that because my mother taught me". "I dislike the other because it caused tension in the family." In terms of this process, both the fact of DI and the identity of the genetic father are likely to be critical information. Whether the question is "where does my sense of humour/musical gift/quick temper/red hair come from?" or "why does mother go quiet whenever

people say that I'm just like my dad? Does she wish I looked like her?" ... the DI offspring is deliberately deprived of key information.

Intriguingly, the law does allow one situation when the cloak of secrecy can in theory be stripped from a donor. If a child born as a result of the mother receiving donated semen finds that they have a genetic disease that can be traced back to the biological father, then the child can sue that father for damages. It's never happened. But the very fact that the law allows for the situation is in itself an acknowledgement that the biological paternal genetic endowment is very much part of who you become, and that this heritage plays a role in setting the limits of your abilities – and your disabilities.

For David, the discovery that his family roots were more complex than he had originally been led to believe had another interesting outcome. Having been deeply troubled by his partial loss of identity he started searching for his father. It is a search that is blocked in almost every direction by legal bars and officialdom. There has, nevertheless, been one avenue that remains open – there is nothing to stop him from getting together with other DI children and seeing if their genes match. For David, this part of the search has been productive.

I have found two half-siblings: children of the same donor father but of a different mother. We do not know who our donor was, although we are committed to trying to find out. But what doubts I have had about the profound significance of blood ties, their importance in the storytelling that is as much the stuff of life as the events it recounts, are gone now. We are three people in middle age. We have, between us, four children of our own, and successful, prosperous professional lives. And yet for me the discovery of these two has been like the warming of frozen soil after a hard winter, bringing growth and green into places that have seemed dead and desolate for 30 years and more. For the three of us, the delight we feel in each other's company, and our certainty about the importance of our connection with each other, have a strength, immediacy and simplicity that is like a taste in the mouth: something known before and beyond the need for discussion. This meeting has brought me great joy – and an even greater certainty about the need for an end to secrecy in DI.

Both in the loss of identity that came from realising there was a hole in his family, and in the discovery of a new tranche of half-siblings, David's story displays the way that his sense of self is linked into his biological family.

A passionate campaign

As the new millennium dawned, the whole issue arrived at court in London. The case brought before the court was designed to challenge the UK's law. Legislation enacted in 1990[2] ensures that any man making a donation of semen also supplies minimal information that is kept on file. The law also permits any person over the age of 18 to enquire whether they had originated from the use of donated sperm, and if so to be given a small amount of non-identifying information. Given that many, if not most, nurturing parents avoid telling their children about this part of their origin, this assumes that the person will for some reason suspect that their history is more complex than is the case for most of their friends and go searching for the data.

At the time of the legislation the authorities decided to ignore the idea that a person's genetic heritage is of importance to them, even though it was clear that most people in society placed a value on knowing their family history.

But time has moved on. The chairman of the committee whose report led to the creation of the 1990 act, Baroness Mary Warnock, is one of a number of people who have changed their minds. She has now been swayed by the strength of argument

---

[2] Human Fertilisation and Embryology Act, 1990.

that there are important medical, cultural and social reasons for allowing DI-children to trace their genetic fathers. She also believes that society has moved on to the point that supplies of sperm will not "dry up" if donors are required to reveal their identities.

The current legal challenge involves two people. One is a child who is too young to be named, and the other is 30-year-old PhD student Joanna Rose. Born in England, Rose moved to Australia at the age of 19, and now lives in Brisbane. In a written statement to the High Court in London, Rose said that she feels "that these genetic connections are very important to me, socially, emotionally, medically, and even spiritually". She is eager to be able to access information that will assist her to form

> a fuller sense of self or identity and answer questions that I have been asking for a long time. I am angry that it has been assumed that this would not be the case, and can see no responsible logic for this (given the usual pre-eminence accorded to the rights and welfare of the child), unless it is believed that if we are created artificially we will not have the natural need to know to whom we are related. I feel intense grief and loss, for the fact that I do not know my genetic father and his family.[3]

I called her and spoke over the phone. "I have lost all of my paternal family, present and past and future – my nephews, nieces, uncles, aunts – everybody," she said. At the same time she recognises that other people do tend to live their lives without considering the issue because having family around is just so normal, but that sense of normality emphasises, rather than diminishes the importance of the issue. "I think you live from day to day without considering your eyesight either," she retorted.

> You live with your family and may not consider it, but if one of those family members goes missing you would be

---

[3] Rose and another vs Secretary of State for Health, 26 July 2002.

distraught. What people are applying to us is a type of battery-animal logic – if you are bred for it you don't know anything different. People seem to think that it is OK if you negotiate this situation into the contract of their life.

Rose has found that this fundamental lack of knowledge of her origins has created an ever-growing void. The older she gets, the bigger it becomes. As a child it had less of a tangible effect but as she grew she became consciously aware that, while other people could look at their parents and extended family to help build their own sense of visual identity and personality, she couldn't.

What the three people we have met in this chapter demonstrate is that knowing your paternal genetic family, in addition to your maternal line of descent, adds to your life. It is important not just because these people are probably your father's friends, although that may not be the case, but because they are family.

Rose points out that a child of donor insemination is presumed to form relationships with the non-genetically related social father's family, while having no thought or care about that of their own genetic father.

> There is all sorts of madness to it. Take a family gathering at a wedding. I am supposed to talk to relatives who I have never met before, purely because I am supposed to be their genetic relative, but I am not. And if I were, then I should feel a sense of attachment. It's schizophrenic ... I think that people give a lot of weight to genetic attachment – it's just that they conveniently wipe away bits to suit themselves.

Rose argues that people tend to think that your mind can be "cut and pasted" – that your loyalties and your response to genetic responsibilities and connections can be re-programmed at will. The idea is that you can cut out any attachments that are inconvenient and paste in new ones that feel nice. It is as though a person's mind and their genetic connections are totally malleable. The hope underlying this is that if you explain to the child what happened and familiarise them with the story, then they will erase the biological connections. "It's totally illogical," said Rose with force.

She sees family members as a type of compass, saying that they provide points of reference that help you plot your own journey and life. By the time she hit late-teenage life the only option she felt was to break away to sort through her identity confusion. It was this quest that took her to Australia.

While supporting honesty, Rose questions the belief that just being honest with a child and telling them about their origins is enough to deal with all the problems. For her, just telling a person that they were launched into life by anonymous sperm donation is of little benefit on its own. To make sense of who she is she needs more. In this court case she is asking for non-identifying information. This will at least enable her to understand the source of some of her physical characteristics, and be equipped to give truthful and valid answers to medical questionnaires. Winning the court case would go some way to filling the void.

Christine Whipp asserts that what they need is information that will enable them to identify and meet the man, sit down and talk. "In effect we are saying that losing the benefit of knowing your father and his family is a disability – an identity disability," said Rose. While there are plenty of people with disabilities living fulfilled lives, that is no argument for deliberately building disability into their lives. This lack of access to their personal heritage and family is in some ways equivalent to being born without a leg – something that if deliberately created would cause an outrage.

To be born into an environment where your mother never even touched your father – it wasn't even a one-night-stand – means that she is giving birth to a partial genetic stranger. I am fully aware that one of the most important things I can hand on to my children, if I even have any, is the best father I can find. That is the best thing you can give them. If you love the person you have children with, you will cherish the aspects of him that you see in them, you will understand the aspects that are difficult because you have learnt to love the person first. I think that that is all really important and beautiful, that aspect of human partnership in reproduction.

Having considered all relevant previous legal cases, the judge, the honourable Mr Justice Scott Baker, came to the conclusion that Rose had grounds to challenge the current situation on the basis that it could infringe Article 8 of the European Convention on Human Rights as incorporated into English law by the Human Rights Act 1998. Article 8 deals with a person's right to privacy, and the judge was indicating that there are grounds for questioning whether withholding information of a person's genetic background prevents them living a full and healthy private life.

From the evidence he collected, the judge drew up a series of principles. While deeming family life to be flexible and elastic, the state has an obligation to protect individuals from unnecessary interference by a public authority. However, he continued, "respect for private and family life requires that everyone should be able to establish details of their identity as individual human beings. This includes their origins and the opportunity to understand them. It also embraces their physical and social identity and psychological integrity."[4]

As a consequence he indicated that the European human rights legislation has an interest in a person's ability to get hold of both identifying and non-identifying information about any donor who brought them into existence. "It is to my mind entirely understandable that [children born from donor insemination] should wish to know about their origins and in particular to learn what they can about their biological father, or in the case of egg donation, their biological mother," wrote the judge.

In some instances, as in the case of the Claimant Joanna Rose, the information will be of massive importance. I do

[4] Rose and another vs Secretary of State for Health, 26 July 2002, para 45.

not find this at all surprising bearing in mind the lessons that have been learnt from adoption. A human being is a human being whatever the circumstances of his conception and [a child born following donor insemination] is entitled to establish a picture of his identity as much as anyone else.[5]

The judge's comments about egg donation raise issues within the fertility industry, as new techniques are being developed to find sources for human eggs. Any techniques that could enable a woman to use eggs generated by her own ovaries, or tissue taken from her ovaries, would help alleviate this problem. At the other extreme, however, would be the occasion when a whole embryo is donated. In this case it is not just the paternal line that is severed, but both roots of genetic inheritance. The person would be left with two channels of enquiry to pursue if they wanted to find the genetic basis to their existence. Cloning, if and when it occurs in humans, will introduce a new twist to this issue. A clone's genetic heritage will be still more confused than a child born after either egg, sperm or both has been donated by an unknown party. For example, the mother could be the clone's twin sister.

This disconnection could take another bizarre twist. Biochemical researchers claim to be close to creating eggs and sperm by carefully culturing normal cells taken from very young embryos.[6] The team of scientists working at the University of Pennsylvania took cells from young mouse embryos that theoretically had the potential of developing into any type of cell in the body; that includes sperm or eggs. They then grew them so that many cells were packed together on a culture plate. In these conditions, some cells form floating clusters that break away from the rest of the mass. Usually scientists discard these cells, but the Pennsylvania team found that these cells could develop into immature eggs if placed in fresh culture media.[7] Another team at the Mitsubishi

[5] Rose and another vs Secretary of State for Health, 26 July 2002, para 47.

[6] Pagan S. (2003), "The next IVF revolution?" *New Scientist*, 10 May.

[7] *Science*, DOI: 10.1126/science.1083452

Kasei Institute of Life Sciences in Tokyo has performed similar manoeuvres to turn embryonic cells into sperm cells. If it can happen with mouse cells, there is a strong possibility that human cells could do this as well. If the techniques are as good as the creators claim, the world could have a source of manufactured gametes – and would be on the way to generate a new genre of relationship-free embryos. It's not a happy prospect.

"Respect for private and family life has been interpreted by the European Court to incorporate the concept of personal identity ... That to my mind, plainly includes the right to obtain information about a biological parent who will inevitably have contributed to the identity of his child," the judge continued. In the convoluted approach of the legal system, he was acknowledging that part of a person's being is their genetic family heritage. The case grinds on.

I do not intend this chapter to be seen as part of the campaign to change the law, though looking at the situation as it exists, and seeing the outcome it has on individuals involved, surely calls into question the practice of donating sperm and eggs, and certainly questions donor anonymity. Instead, I believe that these three people's stories display the way that family relationships – past, present and future – are a vital part of the make-up of each of us. This aspect of what it is to be human is often only apparent when a person's "story" is suddenly stripped away – only then do we come to notice the importance of being a related being.

CHAPTER SIX

a material being

Staring down my A-level microscope, trying to make sense of what it was that distinguished living from non-living systems, I was sadly unaware that I was following a similar pattern of thought to chemist James Lovelock. In 1961 he became involved with a team of scientists at the Jet Propulsion Laboratories (JPL) of the California Institute of Technology in Pasadena. This team was faced with a huge question. What equipment would you need to build into a spacecraft if it was going to detect life on the planet Mars?

What Lovelock didn't realise at the time was that the process of thoughts triggered by this question would lead him to reappraise the whole nature of life on Earth. It would culminate in a theory that his friend and neighbour, novelist William Golding, recommended he called "Gaia". When I met Lovelock at his home tucked away on the Devon–Cornwall border, Lovelock explained that Gaia theory works on the basis that organisms and their material environment, the soil, rocks, rivers and oceans, evolve as a single coupled system. He is passionate that we should not study geology and biology as two separate disciplines. Understanding how living and inanimate elements of the Earth work together shows how the Earth has developed a sustained self-regulating system that has created a hospitable environment for its current spectrum of living organisms. As such we are more than biological; we are part of the physical material that makes up our planet – we are material beings.

Gaia theory shows that not only am I made of an assortment of different chemicals and a few buckets of water, a material being, but also I am a part of the living system that operates in unity with the material Earth.

Since he first went public in 1979, the theory has always been controversial, but in 2001 over 1,000 delegates meeting in Amsterdam put their signatures to a declaration that "the Earth System behaves as a single, self-regulating system comprised of physical, chemical, biological and human components". It's a watered-down version of Lovelock's grand idea, because it ignores the concept inherent in Gaian theory that the goal of any regulation is to create a habitable planet, but all the same, Lovelock recognises that it is a huge step in his direction.

Some scientists say that Gaian theory is as revolutionary to our understanding of how the Earth operates, and to humanity's place on that Earth, as was Nicolaus Copernicus' sixteenth-century announcement that the Earth rotates around the Sun, as do all other planets in our solar system.[1] In its day, this Copernican revolution caused a complete reappraisal of human understanding. If Gaia theory is correct, then human beings are part of a complex network of material relationships. Human beings are undoubtedly made of the same chemical materials that build the rest of the universe, but Gaia now takes this one step further. It says that human beings are part of a huge material and biological mass that makes up the Earth. We are not just made of material, we are an integral part of the material world.

Musings on Mars

Lovelock had been invited to join the JPL team because, by the age of 41, he had gained a reputation as a capable inventor and the American team needed people who could dream up highly sensitive measuring devices and find ways of fitting them into their spacecraft. Consequently, over the next 12 years he made a series of short visits to JPL, and on one occasion was invited to sit in on a meeting that was planning the life detectors that would go on a space vehicle that they hoped would visit Mars.

The key question on everyone's minds was: Would they find life? He sat in a mixture of disbelief and depression as people proposed machines that could attempt to culture micro-organisms from samples of Martian soil, or analyse the soil for traces of

---

[1] Copernicus's ideas were first fully described in his book *De revolutionibus orbium coelestium* (The revolution of the heavenly spheres) which was not published until 1543, even though he had finished writing it in 1530. He was aware that it was a highly controversial theory, and that announcing it could be seen as heresy – punishable by death.

life. "The flaw in their thinking," Lovelock told me, "is their assumption that they already knew what Martian life was like."[2] He was frustrated by the narrow-minded approach that started by assuming that life would have to look like life on Earth.

Lovelock found himself asking some rather down-to-earth questions, such as: How can we be sure that any life on Mars will reveal itself to tests that seek to detect the sort of organisms that form life on Earth? But the question didn't stop there. Lovelock soon realised there was a more critical question: "What is life, and how should it be recognised?"[3] His conclusion was that the scientists at JPL needed to look for "life itself", not for living organisms. Individual life forms might be present in ways that are very different to the familiar versions on Earth, but he felt that if "life" were present on a planet, it would leave its mark.

As the meeting drew to an end, he raised his concerns, and was disappointed, though not entirely surprised, to find that his comments were swept to one side. This team needed people who came up with solid ideas, not an eccentric Englishman who was going to throw philosophical spanners in their works. Lovelock soon found out that someone had heard him, because the next day he was called to the office of senior laboratory chief Robert V. Meghreblian.

After prompting, Lovelock explained that he felt the package of experiments they were planning to send to Mars would be a waste of time, and certainly not worth the huge price tag. Instead, he proposed that they rethought the project and, rather than trying to find living things, they looked for the results of living things – they looked for life's fingerprint rather than trying to grab the finger. Lovelock said that as far as he could see they needed to look for signs of "entropy reduction". As the words left his mouth, he saw a sceptical glaze wash over Meghreblian's eyes; entropy is a notoriously slippery subject and the thought of

[2] Lovelock J. (2000), *Homage to Gaia*, Oxford University Press, pp. 243.

[3] Lovelock J. (2000), *Gaia: A New Look at Life on Earth*, Oxford University Press, p. 2.

chancing the success of your prize rocket on it was far from appealing.

Entropy is a measure of the disorganisation of a system. The greater the disorganisation and chaos, the greater its entropy. The concept is that, left on its own, any "system" will become more and more disordered. It is wrapped up in what has become known as the Second Law of Thermodynamics. Take a child's bedroom as a simplistic example. Without a flow of energy into the room, normally in the form of loud and aggrieved parent, it will soon lose any semblance of order. In scientific terms, the entropy of the system increases and is only kept low by some form of deliberate, or at least energy-consuming, action. Lovelock argued that if there was life on a planet there should be signs of surprisingly high levels of order – there should be a reduction in the entropy compared to what would be expected by pure chemical and physical calculations. You might, for example, find large numbers of molecules where the atoms were frequently placed in the same order. The chief scientist was intrigued, but not convinced. He gave Lovelock two days to flesh out his ideas.

Entropy and equilibrium

Forty-eight hours later Lovelock was back in his office with a set of experiments that he thought the team at JPK should consider including. Each was aimed at searching for signs of a greater reduction in the level of entropy on the planet than could be predicted by making basic chemical calculations of the way that its constituent chemicals would react with each other. The most promising approach, he felt, would be to measure the relative proportions of the gases in the Martian atmosphere. An alternative would be to search for unexpectedly organised chemical molecules in the soil that could also give clues that life existed.

Testing the atmosphere had a clear advantage over scrutinising the soil. The atmosphere was likely to be well mixed and therefore homogeneous throughout Mars. Wherever the probe landed it could suck in a sample and come up with some figures. Testing the soil, if there was any, was more hit and miss. You might land in the Martian equivalent of the Sahara Desert or the great salt pans of North America and consequently find nothing – or you could get lucky and land in one of the relatively small proportion of places on Earth where life flourishes.

With such a short time to pull his ideas together, his comments were necessarily vague, and his visit to JPL was scheduled to come to an end that afternoon. Lovelock returned to his home in Wiltshire, where as an independent scientist he pursued his thinking without the restrictions imposed by a huge organisation. He was excited by the idea of looking for signatures of life in the atmosphere and soil, and set about looking for evidence to support his suggested experiments. Lovelock's first discovery was that there was little to discover – at least, there was very little that he could read on the subject. "I had expected to discover somewhere in the scientific literature a comprehensive definition of life as a physical process, on which one could base the design of life-detection experiments, but I was surprised to find how little had been written about the nature of life itself," he commented.[4]

It's not that people had failed to study living things. There were plenty of data on anything he wanted to know about the beasts that creep and crawl and the plants that reach for the sky. But they didn't answer his question.

At best the literature read like a collection of expert reports, as if a group of scientists from another world had taken a television receiver home with them and had reported on it. The chemist said it was made of wood, glass and metal. The physicist said it radiated heat and light. The engineer said the supporting wheels were too small and in the wrong place for it to run smoothly on a flat surface. But nobody said what it was.[5]

[4] Lovelock, *Gaia*, p. 3

[5] Lovelock, *Gaia*, p. 3.

Lovelock's proposed experiments took a step back from studying the organisms themselves, and instead asked questions about what you would expect to find if there were living organisms around. His conclusion drew on his background in chemistry, where he knew that chemical systems left on their own will move to a point of equilibrium. It's the basic rule behind any simple schoolroom chemistry lesson. On the other hand, if there are living things, plants and animals, then the system is constantly messed around.

In the mid-1970s the idea was viewed with distinct caution, particularly by many of the space-exploring colleagues at JPL. There were some, however, who took his ideas seriously and Lovelock found himself elevated to the position of chief scientist for the physical life-detection experiments on *Voyager*, the next probe due to be launched towards Mars. Within the American team the general feeling was that life was out there ready to be discovered, and their resolve to find it was not even dampened by images that arrived on Bastille Day 1965. The pictures came from *Mariner*, a spacecraft that was orbiting Mars, and they clearly showed that the Martian landscape was barren and rocky. It only served to increase the feeling that finding Martian life was going to be a more exciting challenge than they had originally anticipated.

Then in September that year, Lovelock was told that the funding for *Voyager* had been cancelled. Instead, the space planners were proposing a different vehicle called *Viking* that would go back to the idea of looking for traditional signs of life, such as microbes that would grow in Petri dishes. His dream of seeing a probe that would look for the physical fingerprint left by living organisms was dashed. But all was not lost. In the same month Lovelock was shown infrared spectrum data from the Pic de Midi Observatory in France. These gave a detailed analysis of the chemical composition of Mars' atmosphere. They also had similar data for Venus. "As I suspected, the cratered, moonlike Mars had an atmosphere close to chemical equilibrium, and it was profoundly different from the rich and anomalous atmosphere of

Earth," he explained.[6] The gases around Mars existed in a simple equilibrium with themselves and the rocks making up the surface. The atmosphere was mainly carbon dioxide and bore no signs of the disturbance seen on Earth – if Mars ever had had life it was long-since gone. His conclusion didn't go down well with people dedicating themselves to building a probe that planned to fly to the planet and prove that life existed in Martian soil.

It was a critical afternoon for Lovelock. "Now that I knew the composition of the Martian atmosphere was so different from that of our own, my mind filled with wonderings about the nature of the Earth," he recalled.[7] He pondered the presence of oxygen in the atmosphere, and the fact that it needed to be held within a remarkably narrow range of concentrations over millions and millions of years. "It came to me suddenly, just like a flash of enlightenment, that to persist and keep stable, something must be regulating the atmosphere and so keeping it at its constant composition," he says. "Moreover, if most of the gases came from living organisms, the life at the surface must be doing the regulation ... suddenly the image of the Earth behaving as a living organism able to regulate its temperature and chemistry at a comfortable steady state emerged in my mind." Lovelock saw the complex atmosphere less as a "sign of life" and more as "life itself". On an afternoon's walk up the village street of Bowerchalk in Wiltshire in 1971, William Golding suggested that he called the idea Gaia after the Greek goddess of the Earth. And so his theory gained a name.

An integrated atmosphere

Gaia was no momentary knee-jerk reaction against having his plans for the Mars mission snubbed. By now Lovelock had had time to think through the issues and collect data. Much of his work focused on the composition of the atmosphere on Earth, in

[6] Lovelock, *Homage to Gaia*, p. 252.

[7] Lovelock, *Homage to Gaia*, p. 253.

particular to the co-existence of gases that you would expect to react rapidly, if not violently, together. He was fascinated by the co-existence of oxygen and methane. "The simultaneous presence of oxygen and things that react with it are critically exciting; that is what gives it the disequilibrium," Lovelock told me.

> There is just no way that any inorganic process could lead to both oxygen and methane co-existing because they react quite quickly high in the atmosphere. They are in fact burning in a cool flame in the sunlight. Something has to be pouring them in all the time, otherwise they would both be used up and the flame would go out.

Lovelock points out that plants take in carbon dioxide and, using energy from sunlight, release oxygen. When they die and decompose, bacteria in the soil return some of carbon to the atmosphere, this time in the form of methane. If you want to see this methane, just stab a stick into the mud at the bottom of a pond and watch as bubbles of the gas fizz to the surface. This methane from dead matter and oxygen from living plants pushes the composition of gasses in the atmosphere far away from the equilibrium position that would be expected from a simple analysis of the chemicals on the planet.

Indeed, Lovelock soon realised that any quick analysis of the atmosphere around the Earth showed that its composition of gases was far from equilibrium. The entropy of the atmosphere was far too low for any simple physical explanation. Something was working within the system and that work was having a tangible effect on the system itself. That something, concluded Lovelock, was life. Living organisms were operating in such a way that they pushed the balance way off centre. This is in effect an increase in the order of the system or, to put it another way, is a decrease in entropy, the very fingerprint of life that was absent from Mars.

In an equilibrium world, chemical science predicts that given the chemical make-up of the Earth, you should find an atmosphere of 98 per cent carbon dioxide, with 1 per cent nitrogen, 1 per cent argon and no oxygen. The figures for the atmosphere around the Earth are strikingly different – with 0.03 per cent carbon dioxide, 78 per cent nitrogen, 1 per cent argon and 21 per cent oxygen.[8] Lovelock argues that you might expect to see a tiny amount of oxygen in the atmosphere due to the energy from sunlight splitting the occasional water molecule into its constituent parts, oxygen and hydrogen. The hydrogen could then fly off into space, making it impossible for the atoms to recombine and leaving the oxygen behind. But, he continues, that would be of little use as far as life on Earth is concerned, because you need a reasonably high concentration of oxygen to enable energy-releasing reactions to occur.

The value of 21 per cent for oxygen is just about right. Much less and there would be a large chemical obstacle to these reactions; much more and fires would break out spontaneously. "At 25 per cent oxygen, even damp vegetation will continue to burn once combustion has started, so that a forest fire started by a lightning flash would burn fiercely until all combustible material was consumed," Lovelock claims.[9] This was confirmed experimentally by Andrew Watson, a colleague of Lovelock's, who was working at the University of Reading, England.

> Watson showed that fires cannot be started, even with dry twigs, when oxygen is below 15 percent; above 25 percent oxygen, fires are so fierce that even the damp wood of a tropical rain forest would burn in an awesome conflagration. Below 15 percent there could be no charcoal; above 25 percent no forests.[10]

The chance of this happening by physical processes is infinitesimally small. "If you try to calculate the odds of getting the co-existence of oxygen and methane, not even bothering

[8] Lovelock, *Gaia*, p. 33.

[9] Lovelock, *Gaia*, p. 34.

[10] Lovelock J. (2000), *The Ages of Gaia: A Biography of Our Living Earth*, Oxford University Press, p. 124.

about the other gases, in an atmosphere where there was no life, the computer packs in at its limit of precision – it's just not possible", Lovelock explained to me.

Lovelock likens the situation to walking out onto a freshly cleansed sandy beach just as the tide is pulling back. The neatly rippled sheet, however, is disturbed in one place by a carefully constructed sandcastle. On a planet that has a constant input of energy in the form of sunlight, he argues that you will never see the featureless neutrality of complete equilibrium. Instead, you could expect a lifeless steady state where in effect tides keep the beach clean and tidy. The presence of the sandcastle, this perturbation in the system, points to something that is actively disturbing this status quo, and if you catch a glimpse of the human who built it, you will see the agent and, says Lovelock, "The constructions made by the living creature contain a wealth of information about the needs and intentions of their builder."[11]

Lovelock's main thesis is that this stable state of order within the atmosphere was as staggeringly unlikely to have happened without life being present, as was a sandcastle to appear spontaneously as a tide pulls back from a beach. "Sandcastles would vanish from the Earth in a day if there were no children to build them," says Lovelock, continuing that if life were extinguished, the available free energy for lighting fires would similarly disappear. In this case it would be over a matter of a million years or so, but then that is little more than a sweep of the tide in cosmic timescales. He published his ideas in a paper in the internationally renowned journal, *Nature*.[12]

For an example of the way that living organisms alter the atmosphere, Lovelock points to geological features like the White Cliffs of Dover. "Those are great piles of organisms, nothing but a pile of skeletons – a thousand feet of it. And think how much carbon dioxide that represents. If all that gas was in the

[11] Lovelock, *Gaia*, p. 31.

[12] Lovelock J. (1965), "The physical basis for life-detection experiments", *Nature*, 207, p. 4997, 7 August.

atmosphere we would be like Venus," he said. Carbon dioxide acts like the glass in a greenhouse; it allows energy from the sun to enter a planet's atmosphere so that it can heat the surface but the gas prevents heat radiating back out into space. The more carbon dioxide, the hotter a planet becomes. With 96 per cent of carbon dioxide in its atmosphere, Venus has a surface temperature of a scorching 482 degrees Celsius; far too hot to sustain life.

Co-existence and control

In the decades since he first announced his idea, Gaia has been praised and denigrated with about equal force on each side. Part of this is due to the misunderstanding that has arisen from the name, because Lovelock is adamant he is not proposing that Gaia is a self-aware goddess, with foresight and feeling. Instead, he sees her as a massively complex self-regulating mechanism that pools the resources of life on Earth so that they operate in a massive game of "scratch my back and I'll scratch yours". And yes, he frequently uses the female pronoun, defending it on the basis that sailors use it for their boats. No sailor really believes that his ship is alive, but having clung to the deck of a small boat midway across the English Channel in storm-force winds, I can assure you that you soon realise your life depends on this inanimate object and it is easy for the relationship to become personal.

As a reflection of his sense of humour, Lovelock has a statue of a Greek goddess in a flowerbed just outside his home. "There she is," he said with a smile.

But a critical aspect of Gaia is the realisation that the relationship all living organisms have with the Earth becomes more than simply personal, it is intimate. And human beings are not exempt from it. We, like all living things, not only use the atmosphere that surrounds us and the environment we move through – we create it. We are an integral part of the material fabric of planet Earth. According to Gaia theory, there would be no life-supporting atmosphere without living organisms. More than this, the theory says that these organisms and their environment don't just exist alongside each other, but that a powerful interaction exists that keeps everything on Earth co-existing in a self-sustaining symbiosis.

Gaia research, maintains Lovelock, is simply another branch of mainstream science. It draws together biologists in investigations of the way that living organisms help each other out, their study of symbiosis; physiologists, in their quest to understand the way that organ systems operate in the body, and their investigation of the regulatory control systems that keep life and limb together; and geologists trying to understand how the material Earth evolved.

As an example of a part of the Gaian system, Lovelock likes to point to the supply of iodine. In mammals the thyroid gland in the neck scavenges iodine from the tiny levels in which it is found in food and water. It does this because iodine is needed in some of the controlling hormones that flow around the body. The question is: How do we get a regular low level of iodine from our environment? His answer starts in the seas. Long straps of seaweed, laminaria, inhabit inshore waters.[13] They are covered with water while the tide is in, but exposed to the air when it retreats. One of the things they do is to take in iodine from seawater and turn it into a set of iodine-bearing substances, one of which is methyl iodide. This is highly volatile, boiling at temperatures as low as 42 degrees Celsius. As the tide pulls back, the seaweed lies exposed on a beach and this methyl iodide escapes into the atmosphere and, in a matter of hours, exposure to sunlight causes it to decompose and release iodine. This too is sufficiently volatile to stay in the air and be blown hundreds of miles inland, right to the centre of continents. It is then available for us to ingest and use in our bodies, although people living in the Himalayas can develop goitres in their necks as the thyroid gland expands because iodine is in such short supply in these

[13] Lovelock, *Gaia*, p. 110.

land-locked mountains. For Lovelock, the shoreline seaweed is in effect Gaia's thyroid, harvesting iodine from the sea and making it available to the rest of her "body" – part of which is you and me.

"If you think about it, seaweed gets its reward, because animals excrete nutrients, urine and the rest of it, that eventually wash down the rivers into the sea to feed seaweed," laughed Lovelock.

There is no evidence that the supply of iodine to land-based animals increases if demand goes up, but many aspects of life on Earth appear to be held within critical limits. The amount of oxygen in the atmosphere, or the average global temperature are two key examples. There is evidence that not only does life create its own environment, but it does so in a highly controlled manner. Control systems are now so much part of our everyday lives that we often fail to notice them. Turn on your oven and a sensor ensures that the temperature inside stays close to the setting on the dial. When it gets too hot, the energy flow, either in terms of gas or electricity, is automatically reduced. As it gets cooler, the energy is turned back up. Consequently the temperature in the oven "hunts" around the level set in the recipe. Rooms that need to be held at very specific temperatures also have cooling systems and the balance between heaters and air-conditioners creates stability.

Within a body, hormones perform similar tasks. Insulin and glucagon work against each other to maintain incredibly stable levels of sugar in the blood. Insulin appears in the bloodstream when there is too much sugar, causing the body to mop up the excess, and glucagon appears when sugar levels drop, this time causing the breakdown of energy stores so that glucose can flood into the blood and restore the level. Additionally, a complex set of hormones enables women to release an egg and become pregnant – assuming that is that she has intercourse at the appropriate time.

As an inventor, Lovelock was used to the idea of using feedback control systems, and was not surprised to see them in living animals. The question before him then was whether there were any feedback mechanisms operating at a global level. Could, for example, there be some system that maintained the temperature of the Earth at a constant level throughout millions of years? Could the Earth have a way of protecting its atmosphere and ensuring that it retained about 21 per cent of oxygen? Lovelock believes the answer to all of these questions is "Yes".

In the early 1980s Lovelock started to see if he could build a computer simulation of a model world, to assess whether Gaia, if she existed, could maintain a stable temperature. He stripped the world down to be as simple as possible and populated it with just two species of plant. Both were daisies. One had dark flowers and the other had white ones. Without "trying to do anything" these would operate in a way that is analogous to insulin and glucagon. The dark flowers would absorb heat and raise the planet's temperature and so act as an equivalent of insulin, while the white flowers operate in the opposite direction, radiating heat and cooling down an over-heating planet, and as such are the equivalent of glucagon. He set this imaginary world orbiting around a sun that was to all intents and purposes much like our own – it increased the amount of heat it gave off as it got older. Then he sat back and watched.

> The model showed that the natural selection of daisy species growing on this planet led to the self-regulation of climate at a temperature near optimal for plant growth, despite large variations of heat. When the star was young and cool, dark daisies covered the planet and, by absorbing sunlight, made it 17 °C, warmer than it would have been without them. As the star warmed, the lighter daisies began to grow and compete, and their reflection of sunlight cooled the planet and kept the temperature optimal as the star increased its output of heat.[14]

Since then various scientists have built more complex Daisyworlds, and some now have ecosystems with up to 30 coloured plants,

[14] Lovelock, *Homage to Gaia*, p. 264.

12 types of plant-eating animals and three species of carnivore. One intriguing feature is that if the temperature remains too stable, then species start to die out, and in the end the planet is left predominantly populated with only two or three. Introduce a change, the equivalent to the arrival or retreat of an Earthly ice age, and there is an explosion in diversity. "If this view is right then species richness is a symptom of sudden change during a state of health," concludes Lovelock.[15]

"People who don't understand Gaia always think it should be perfect. Many of them say to me, 'It can't be a very good system. Look at the climate, it fluctuates between ice ages and interglacials. Pretty rotten thermostat. No engineer would have made anything that bad'," Lovelock explained to me.

> My response is to say, "Consider yourself." Nobody would doubt that the human body has a stable temperature, yet all humans vary their temperature between 35 and 40 Celsius – a five-degree range. That is almost exactly the same as the earth does. In an ice age, the "core" temperature is 11 degrees Celsius, and now it is 15 to 16 degrees Celsius. We and the Earth regulate at all that is needed – we don't try and go for perfection.

Faced with criticism that Daisyworld only worked because it was designed to work this way, Lovelock took it one step further. This time he populated his model with grey flowers, but gave them the option of mutating their colour – in effect they could gently change their genetic constitution. As the sun cooled, his pseudo-world filled with darker flowers, and the temperature on Daisyworld remained stable. Letting the sun get warmer caused the population to move towards almost pure white flowers.

Developing models is one thing, but Lovelock and key colleagues such as American microbiologist Lynn Margolis are now convinced that they have shown how similar mechanisms can create and maintain stable levels of carbon dioxide, oxygen

[15] Lovelock, *The Ages of Gaia*, p. 215.

and nitrogen in the atmosphere. There is evidence that ocean algae, dimethyl sulphide gas, clouds and climate are all linked together. In all of these, the biological characteristics of living organisms work together to provide themselves with the world they wish to inhabit.

Throwing a human into the works

The key question then, as far as this book is concerned, is: What does a Gaian view of existence say of the role of human beings in the world? Lovelock maintains that human beings have evolved within the living system, and form a part of Gaia's global structure. They use resources and create waste, and very often Gaia can adjust the way she operates to absorb the impact. But in his assessment, it is Gaia's goal to have a living system on Earth, not to support human beings per se. The difference is critically important.

One general implication of this for human beings is that if we get too casual in our relationship with the rest of the world, then Gaia might just have to remove us, or at least reduce our numbers so severely that our destructive impact is minimised.

> There is a natural consequence of a self-regulating system. If any organism goes and spoils the environment in its locality, the chances for its progeny are diminished. It is like dark daisies in a hot world. They are not going to flourish, because the chances that they will grow and produce seeds are greatly diminished. And so anything that has an adverse effect on its world will in effect be "punished". There is no moral force exerting the punishment, it is just a feedback consequence.

"It was Vaslav Havel who pointed out that Gaia was the first scientific theory that had an ethical content, it gave us something to whom you are accountable." Lovelock paused. "That was quite a discovery – but then he was a remarkable man."

In *Homage to Gaia*, Lovelock describes his reaction to a talk given by Mother Teresa, who chastised her Oxford audience over their "concern for the Earth". "We needed," she said, "to take care of the poor, the sick and the hungry and leave God to take care of the Earth." Lovelock waited for the speech to end and then rose to reply: "I must disagree with the reverend lady. If we as people do not respect and take care of the Earth, we can be sure that the Earth, in the role of Gaia, will take care of us and, if necessary eliminate us."[16] Reflecting on the incident, Lovelock was keen to emphasise that he respected her when she was on her subject, but thought it was unfortunate that she had strayed from her subject.

In his book *The Ages of Gaia*, Lovelock expands the grounds for his concern:

> Gaia, as I see her, is no doting mother tolerant of misdemeanours, nor is she some fragile and delicate damsel in danger from brutal mankind. She is stern and tough, always keeping the world warm and comfortable for those who obey the rules, but ruthless in her destruction of those who transgress. Her unconscious goal is a planet fit for life. If humans stand in the way of this, we shall be eliminated with as little pity as would be shown by the micro-brain of an intercontinental ballistic nuclear missile in full flight to its target.[17]

We have a tendency to see ourselves as the pinnacle of biological life, but even there Lovelock sounds a note of caution. "What about whales? Have you ever seen specimens of a large whale's brain? It's huge!" he asked me, adding that large whales have brains 10 times the size of the human brain.

> Now nothing in biology is redundant. As soon as you stop using it, it withers away – use it or lose it. So what is the

[16] Lovelock, *Homage to Gaia*, p. 376.
[17] Lovelock, *The Ages of Gaia*, p. 199.

whale doing with ten times as many brain cells as we have got? Comparing all other animals, brain size and intelligence go roughly in proportion. There is that huge organ – what the hell is it doing? The mere fact that they are not aggressive and don't fight back means nothing.

In addition, everyone knows that our interaction with the Earth is frequently not particularly pretty. "Cutting down trees to build houses and liberate fuel is something that Gaia can probably cope with," said Lovelock. "But obliterating vast tracts of rainforest and leaving desert-like swathes of wasteland behind could so destabilise the environment that Gaia may well let a climate of pestilence strike a blow at the perpetrating species – at us."

He reached across his desk to grab a recent copy of the science journal *Nature*. In it there was a report showing that the moment you start to clear areas of tropical rain forest, you increase the risk of the remaining forest being devastated by fire.[18] "It always seems to me that tropical forests are hanging on precariously – they are only there because they sustain their own environment; really there should be deserts in that part of the world. But the more you clear them away the worse the climate gets," commented Lovelock. Clearing areas of forest reduces the overall humidity in the region, so increases the risk of fire. "It only takes one lightening flash – the forests are going faster and faster and faster. Yes, we are a damn nuisance as far as the planet goes."

The paradox according to Lovelock is that humankind could end up being the saviour of Gaia. The aggressive warlike striving for dominance has driven mankind to build all sorts of machines and some of these could prove useful in the long term, the very long term. Over the next billion years the sun will increase its heat output. Human beings are the only species on Earth capable

[18] Cochrane MA. (2003), "Fire science for rainforests", *Nature*, 421 pp. 913–919.

of building a sun-shield in space, a huge parasol that could maintain the temperature. We are also the only species that could devise ballistically armed spaceships that would have the capability of blasting planets that threaten to strike the Earth and as such warding off interplanetary impacts. Both may seem far-fetched, but there are research teams looking into their feasibility.

So much for humankind's place within the Earth. What about me? What does Gaia say to a person's sense of their identity? To start with, Gaia puts humanity in its place. It removes any heady arrogance that says, "I can make it on my own", because fundamentally our very existence is dependent on the physical and biological world. It also, however, shows us how important we are, in that the physical and biological world only exists as it does because we are present within it. This is far more than saying that there would be no hedges or meadows without us, but that every breath I take places carbon dioxide in the air, and as such links me to the vast carbon dioxide deposits locked in chalk downs and seaside cliffs.

Lovelock claims to look at life from an agnostic background with Quaker overtones. "It is very nice to have something big and stern out there that does almost everything that the gods are supposed to do. This I find is very comforting – that I am part of this gigantic system that has existed four billion years," he said and then laughed.

> I find it more intriguing at my age [he was born in 1919]. The system is just a little younger than me. It can't have more than a billion years to go at most, so four-fifths of its life span has gone. This makes it very familiar. I feel that it is an elderly lady who should be treated with respect.

Gaia is still a theory waiting to mass enough evidence to get as close to universal agreement as you could ever hope for in science, but even if the final details of the scheme don't shape up, there is no doubt that it has already altered our understanding of ourselves.

Material by design

Gaian theory isn't the only way of looking at our existence that causes us to recognise that we are material beings. Another comes from a very different viewpoint. The twentieth century saw a furious slanging match between two groups of people. On the one side were evolutionists who were convinced that scientific evidence pointed to a multi-billion-year-old universe and a progressive development of life from primitive single-celled microbes to complex plants and animals. On the other were religiously minded people who confined their thinking to biblical narratives of the origins of life the universe and everything. For them the origin of life was a few-day creative event orchestrated by an all-powerful God that occurred somewhere around 6,000 years ago.

Sat in the middle of this argument was a third group. These people point out that it is perfectly possible to draw the two opponents together while still being true to both. Professor of Old Testament Walter Brueggemann points out that the first few chapters of the biblical book of Genesis are among the most important in biblical scripture but they need to be read with care. "There is no doubt that the text *utilizes older materials*," he claims. "It reflects creation stories and cosmologies of Egypt and Mesopotamia."[19] The biblical writer is then re-engineering these myths in a way that describes an idea that God and humankind are made to work together.

"At the outset," he continues, "we must see that this text is not a scientific description but a theological affirmation. It makes a statement of faith." The story is told in a classic style of Hebrew poetry where a pattern is set up in the first set of verses, which then echoes in the second batch. In this case, day one sees the arrival of light, day two brings sky and in day three we hear of the creation of seas and dry land. Then the pattern repeats. Day four places the sun, moon and stars – the objects that give

[19] Brueggemann W. (1982), *Genesis: A Bible Commentary for Teaching and Preaching*, Atlanta: John Knox Press. p. 24.

light; day five we get creatures teeming in the sea and in the sky and day six comes with the land animals, including human beings.[20] Each "day" is then set out in a remarkably symmetrical structure:

time: "There was evening and morning ... "
    command: "God said, 'Let there be ...' "
        execution: "And it was so ... "
    assessment: "God saw that it was good ... "
time: "There was evening and morning ... "[21]

Of interest for this chapter is that the writer uses the same word, the Hebrew word *bara*, to describe the way that God creates the material universe, as he does for the creation of the first living creatures and of the creation of humankind. Humankind may have distinct aspects of uniqueness in the way they relate to each other and to God, but their origin is much the same as everything else.

To its original readers, this ancient account would have been comforting and affirming because it placed them in a purposefully created structure. Intriguingly though, it is not the only account of human origins given in these scriptures, because no sooner than the seventh day, a day of rest, is out of the way than the author finds another way of telling the story. This version is more shocking. The author could easily have attempted to tell of humanity's origins by saying that we are fallen from stars, or arrived on a meteor. Humans are obviously supreme; they must therefore have had a wonderfully exalted start. But no, when this scripture uses pictorial language to describe our origins it says we are created from the "dust of the earth",[22] from the same material that makes mountains and molehills. Nothing there to make us feel too important about ourselves. In an ultimate way, it puts humankind back in its place; a concept

---

[20] Genesis 1:1–27.

[21] Brueggemann, *Genesis*, p. 30.

[22] Genesis 2:7.

reflected in the familiar words spoken at many a funeral service: "Dust to dust, ashes to ashes".

In terms of our physical and biological origins, this chapter has reviewed two very different ways of looking at the world: modern science and ancient religion. Both, however, point in the same direction. Both are very comfortable with the idea that we are made of the same "stuff" that makes up the rest of the planet, indeed that makes the rest of the universe. As such we are material beings.

CHAPTER SEVEN

a spiritual being

O ne of the most contentious issues relating to any discussion about the nature of human existence and our experience of being an individual member of our species is the notion that we are spiritual beings. The controversy, as we shall see, is less along the lines of discussing whether we are spiritual. Instead, the issue is contentious because there is a wide variety of views as to what spiritual means.

A simplistic common perception in Western European countries is that most people have given up belief in organised religion. It is an easy next step to presume that they have renounced anything associated with a notion of a divine being and have no interest in spirituality. This at first sight appears to be backed up by statistics such as the drop in church attendance in the Church of England from 2.9 million in 1970 to 1.7 million in 1990.[1] Similar, though less severe declines have been seen by other established Christian churches.

While there is certainly growing unease about many forms of institutionalised religion in the UK, this isn't the whole picture, even in the UK. Data from the 2001 census conducted by the Office for National Statistics show that 71.6 per cent of people in the UK still classify themselves as Christian, with only 15.5 per cent saying that they definitely have no religious faith.[2] Move away from asking directly about organised religions and the numbers change again, with some surveys even indicating that interest in spirituality is on the increase. For example, while one opinion poll carried out in 1986 revealed that just under half of the British public claimed to have had a spiritual experience,[3] a similar poll repeated 13 years later recorded that the proportion had risen to three-quarters of the population.[4]

[1] Office for National Statistics, Dataset ST301319.

[2] The data were collected in the 2001 national census and can be found at http://www.statistics.gov.uk

[3] Hay D. and Heald G. (1987), "Religion is good for you", *New Society*, pp. 20–22, 17 April.

[4] A survey to accompany the BBC's *Soul of Britain* series of nine episodes, broadcast in June and July 2000.

In the USA the situation is different. In 1990, 47 per cent of the American population attended church regularly, and by 2000 this had dropped only to 46 per cent. Around 70 per cent of the population claim to be Christian.[5] Clearly on the western side of the Atlantic, spirituality and religion are still closely linked.

One of the key issues is to unravel that link between spirituality and religion. I set out to discover whether the spiritual facet of our being was a tool that enables human beings to seek a god and as such to become involved in a religion, or whether spirituality could be defined in purely secular and humanistic terms.

I wondered whether it was possible to see the relationship between a person's spirituality and their religious outlook as analogous to the relationship between a person's sexuality and their sexual activity. Sexuality influences many different aspects of a person's life and the way that they relate to other people, only one of which is their pattern of sexual activity. People's sexuality is part of the way they are made and also reflects the way they are shaped by their upbringing, but the way that they express that part of their being in sexual intercourse is strongly influenced by moral values, religious teaching and legal rulings of the society they live in. Sexual activity is often the issue that becomes debated endlessly because it is the most obvious outward revelation of a person's sexuality, but to limit the discussion of sexuality to sex would miss the full extent of the issue.

In terms of spirituality, could you say that a person's religious behaviour was the tip of the spiritual iceberg? If so, it would be a mistake to say that the tip was all there was; equally it would be a mistake for religiously minded people to forget the mass of spiritual existence that supports their faith.

Imagination, aesthetics and biology

I was keen to discuss the issue with the philosopher Baroness Mary Warnock. Having spent a lifetime in education she now

[5] http://www.teal.org.uk/stats/key.htm

sits as a life peer in the House of Lords. In one of her recent books she talks of the need to get the "spiritual back into education", and I wanted to explore what she meant.[6] So on an overcast afternoon I headed past the police sentry and on through the ornate entrance of the House of Lords, into the peers lobby – a wonderfully calm foyer that gives you the sense of stepping into a gentlemen's club, which despite attempts to bring about reform, it still is in many ways.

"On the project of what it is to think of oneself as fully human, to be fully human, I think the spiritual dimension is enormously important, but very difficult to pin down ... My general definition of the spiritual dimension of a person's life? It is what really exercises a person's imagination, it would embrace the aesthetic," Warnock told me. As a consequence of her experience within education, she derides the job-driven philosophy underpinning the current UK curriculum. The problem is that it concentrates on teaching facts, and squeezes out any scope for individual discovery. This, she says, removes any chance of introducing children to their spiritual nature.

She is sad that many people think spirituality in schools should be confined to religious education. But even in religious education she sees little if anything that helps pupils' spiritual development because they are in reality lessons in comparative religion – a series of facts, cold assessments and statements about various world faiths. She was not saying that this was in itself wrong, but that she thought it could be so much more.

In my view, religion can't be disentangled from the aesthetic. Therefore one would want to try to educate children so that they at least had some idea about the amazing effect of music, poetry, paintings, things which actually go beyond the capacity of words to express. If I were a teacher in a class of children I would feel I had succeeded if they began to see there is more in their life and their environment than

[6] Warnock M. (1998) *An Intelligent Person's Guide to Ethics*, Duckworth, p. 179.

can be physically seen. There is more in things like playing their instruments and listening to music than can be thoroughly expressed in words or in scientifically true or false accounts.

She believes that it is the spiritual facet of human beings that allows them to see beauty in an arrangement of coloured blobs of paint on a canvas, and to be emotionally moved by an operatic performance; to see the true likeness in a portrait; or feel pain and excitement in music.

The question in my mind then was where this ability to imagine, to feel things that don't exist, comes from. How is it built into human beings?

I found one answer in a book that studied the development of spirituality in children. *The Spirit of the Child*[7] was written by zoologist David Hay and his research associate Rebecca Nye while they were working at the University of Nottingham, England. His research suggests that spiritual awareness is hard-wired into human beings. Why? Because it has survival value. He claims that his research supports "a view of spiritual awareness as a natural human predisposition, often overlaid by cultural construction, but nevertheless a biological reality".[8] He argues that "children's spirituality is rooted in a universal human awareness; that it is really there and not just culturally constructed illusion".[9]

When explaining his understanding of spirituality, he again refers to metaphor, saying that some people talk of a spiritual journey in which their religious views are their mode of transport. Others talk of spirituality as the fuel that enables the vehicle of religion to operate. "It is as if most people, even those who have no time for the religious institution, see the need for some vehicle of spirituality," says Hay.[10]

[7] Hay D. and Nye R. (1998), *The Spirit of the Child*, Fount.

[8] Hay and Nye, *The Spirit of the Child*, p. vii.

[9] Hay and Nye, *The Spirit of the Child*, p. 4.

[10] Hay and Nye, *The Spirit of the Child*, p. 7.

The first person to put forward a biological view of spirituality was zoologist Alister Hardy. In 1965, shortly after he retired from the Chair of Zoology at Oxford University, Hardy gave the Gifford Lectures at the University of Aberdeen. As a committed Darwinist, he proposed that religious experience or, to use the language of this chapter, spiritual awareness, has evolved through the process of natural selection because it has survival value to the individual.[11] Hardy was suggesting that this form of awareness is potentially present in all human beings and has a positive function in enabling individuals to survive in their natural environment. He provided an evolutionary mechanism to explain the biological mode in which spiritual awareness emerged in the human species. "On Hardy's thesis, spirituality is not the exclusive property of any one religion, or for that matter of religion in general," comments Hay.[12]

Relating to religion

"I don't know what this biological substrate is supposed to be ... " was the Archbishop of Canterbury's reply when I mentioned this biological theory to him as we sat in his drawing-room at Lambeth Palace across the river from the House of Lords. Rowan Williams had recently been enthroned as the leader of the worldwide Anglican church, but in the run-up to his election he had caused a storm by being inducted to an ancient Welsh order of druids. The photos of his dishevelled hair, wild eyebrows and bearded face framed in the white hood of the ceremonial robes made for great front-page pictures in national papers.[13] During the ceremony Williams was given the bardic name of *ap Aneurin* and afterwards described the award as "one of the greatest honours that Wales can bestow on her citizens".

---

[11] Hardy A. (1966), *The Divine flame: An Essay Towards a Natural History of Religion*, London: Collins.

[12] Hay and Nye, *The Spirit of the Child*, p. 11.

[13] Savill R. (2002), "Archbishop in waiting becomes druid", *Daily Telegraph*, 6 August.

In justifying his action he pointed out that the circle of druids he had joined was more akin to a Welsh-language poetry-reading group than a society of pagan worshippers. All the same, the event gave rise to complaints that he might have a curiously broad notion of spirituality for an archbishop.

"I don't know what this biological substrate is supposed to be that enables imagination, nor does anyone. I have to say I am a bit sceptical about it – I don't know how we would begin to test for that. I know people say that contemplative monks live longer – that to my mind faintly trivialises the whole business," he said with a laugh, pointing out that they don't go through the stress of commuting, or chairing committee meetings, all of which could have a greater impact on their longevity.

I like to think that the habit of silent prayer gives you at least some cushioning against some of the more acute and uninterrupted forms of stress, but I would never recommend to anybody that they undertake contemplative prayer to avoid stress. Some of the great contemplative saints have not lived exactly stress-free lives. Of course, if you open yourself up to God, God knows what will happen. It could leave you a lot more worried than you were to start with.

The trouble is that some of the neo-Darwinist language about survival value is so – well, to put it bluntly – is so ropy philosophically, that I wouldn't want to hitch my wagon to that.

I asked for his definition of the spiritual part of human beings. Before answering he politely rearranged my question. He was worried by my phrase "the spiritual part of humans". "It suggests there is a kind of territory in humanity that you mark up – there's the physical bit, there's the spiritual bit," he said, continuing that this division was as meaningless as blandly studying a painting by analysing the paint and the canvas.

But you asked what my definition of spiritual was – it is that dimension of our life lived in relation to God, and as a result of that, lived in a certain kind of relation to others. This includes a whole range of other things, and enables a whole range of other things. I don't think it will do to define the spiritual as the non-physical because spiritual life is physical life lived in a certain way. Even in the Christian tradition where you believe in life after death there is the resurrection of the body, not just the immortality of the soul.

Williams was comfortable with the idea that everyone has a spiritual dimension, but recognised that there are many people who don't realise the full implication of this facet of their existence to the point that some deny it, although this, he felt, was as implausible as denying that you have a physical dimension to life. As far as his experience is concerned, however, Williams says that he has never known a time when religion wasn't part of his life, and as such he feels he grew progressively to understand the spiritual element of his nature.

He told me that since his early memories he had always attended church with conviction and grew in his belief in the Christian faith, but he was unsure whether he would describe his early perception of Christian worship as a spiritual experience.

Spiritual experience, I would say in the most general sense, is an experience that projects you into another level of, not exactly understanding, but apprehending, or sensing, the whole of your environment. Therefore one of the characteristics of a spiritual experience is often that you are left not quite knowing how to talk about it; looking for new words about it. That doesn't mean it has to be an experience of mystical ineffability, just that there is something more going on than you ever realised. I think for a Christian, the kind of trigger and the kind of newness involved would be perhaps a sense of being seen, loved and held by God – by an agency that is not identical with any bit of the world around you. It is more comprehensive. And I suppose such early experiences as I had of that were mostly in the context of public worship – it's not fashionable right now, but sacramental worship – that sense of belonging in a much bigger world than there was there and being in the light of God, in the vision of God, in a way which left me very much, still leaves me very much at a loss for words. And I guess in my teens I would want to say that there were moments when that spilled over increasingly into my private prayers. That sense of excess, being at a loss to describe the experience.

"How much is this sort of spiritual experience an everyday occurrence, and how much an occasional event?" I asked.

He explained that spiritual events often projected you into new experiences, which could be disturbing and exhausting and as such you could expect that to occur each day.

We'd collapse – and of course people who go through a very intense short period in which a lot of that does happen often do collapse. It's a strain to mind and body. And when some of the great mystical writers talk about moments or periods of rapid movement and discovery, they do talk about near collapse in mental and physical ways, it's like a breakdown. But for most of us it will be a moment here or there and then a plateau and then maybe, if we keep ourselves open, another trigger, another projection. With some people it is more dramatic than others.

"Let me give one example from history – the autobiography of a seventeenth-century Welsh monk, Augustin Baker, from Abergavenny, a local man," he grinned, alluding to the fact that he had spent his previous years in South Wales.

He talks about his time as a novice when over a period of three or four days he felt his prayers suddenly going in all sorts of different directions, and new horizons opening up, and he describes the physical impact of it, shaking and twitching and deep physical discomfort and not being able to sleep and that sort of thing. And then it settles down. St Teresa talks about a similar sort of thing. As if the system is receiving more signals than it is used to taking – it is a sort of overload.

Implicit structure

A week earlier I had met another clergyman who was less certain that spirituality could be defined so narrowly. "If you have defined it, you must be talking about something else," commented Edward Bailey in his church office in Winterborne, just to the north-east of Bristol. "Spirituality can't be defined, but it can be pointed to." Bailey has made a career of studying what he calls implicit religion, the idea that we belong to groupings that demand specific allegiances from us. These could include our family, work or sports club. Indeed, by his reckoning, most individuals are members of a number of different implicit religions, each pulling us in different directions.

I mentioned my forthcoming visit to Lambeth Palace.

One of the interesting things about Rowan Williams is that you can see something that is meant by spirituality. I think you can see it by just looking at a photograph of him. You have only got to look at a person like that, or indeed Mother Teresa [of Calcutta]. It is not *simply* physique, it is not *simply* intelligence, it is not *simply* education, it is not *simply* social or cultural, although all those things apply. But there is something here to which we can give the word "spiritual".

So for Bailey, the first thing to note about spirituality is that while it can't be defined, you can point towards it.

Then you could start analysing that little extra at the end of the continuum. I would say that spirituality transcends all those other things, and it is above all about relationship. I am reminded of the genius of Princess Di. You can't define it but you can point to it. Her gift was the ability to enter into relationships. She related to people and they identified with her.

Second, he continued, not everyone has religion, but everyone has a spiritual dimension, just as everyone has a physical dimension. But in claiming that everyone has a spiritual dimension he stressed to me that he was not making any form of value judgement. "You could have an evil spirituality, as well as a good spirituality. Just as you can have a diseased body as well as a healthy body. On its own the term is totally neutral." In a similar way he was at pains to point out that not all religions are spiritual.

"African religion, as I understand it, claims that different people have different amounts of spirit and that your amount of spirit can vary from day to day," said Bailey. Once it was pointed out it seemed a very obvious statement.

After all, we all know that on Monday morning we have a Monday morning feeling [our spirit is low] and on Friday evening we have a Friday evening feeling [we may be physically tired, but the spirit is high]. When we look at Tony Blair winning the general election in 1997 we might say that he is charismatic, he has more spirit than poor old John Major who was on his way out. He was low – he was having a bad hair day.

Islamic spirituality

I wanted to see how the idea of being spiritual was lived out in a closely regulated religion so I met up with Dr A. Majid Katme, a retired psychiatrist who was born in Lebanon, studied medicine in Cairo and came to Britain 30 years ago. He is also a practising Muslim and a vociferous campaigner for the rights of unborn children.

"I discovered Islam while at school in Lebanon. My parents were not religious or practising Muslims, but I attended an Islamic circle that taught Islam. I was so happy to discover our holy book *Al Qur'an* and its wonderful meanings. I loved the guidance to our life with the organised spiritual nourishment found in the five daily prayers," he said, explaining that the prayers follow a strict pattern that involves ritual washing, standing, kneeling, bowing, prostration and readings from the Qur'an.

> When I am kneeling I say, "Praise be my Lord, the Greatest." When I prostrate on the floor I say, "Praise be my Lord, the most High." And I finish my prayer by saying, "*Assalamu Alaikum*, peace be with you." The result is that you have a peaceful and tranquil personality.
>
> With the Qur'an, there are also the *Ahadith/Sunnah*, the sayings and the practice of the final prophet Muhammad, peace be upon him, which were printed in about 20 volumes. Taken together, these two Islamic references give us all guidance and the answers to all questions and issues in life. When I discovered all this I said, "Oh my God, this is what I want, this is what I am looking for, I feel lost without it."

By learning and studying Islam and practising it sincerely and daily, Katme claims that he gets closer to God with an amazing "sweet taste of spirituality".

> In Islam, spirituality is linked to the soul, which is called *ruh* in the Qur'an. This is breathed from God into the fetus at about six weeks of pregnancy, and after that every human being becomes spiritual. Also in Islam, spirituality is translated as *ruhiah* in Arabic when it is normally linked to a person in prayer or meditation who has a peaceful and tranquil psyche and personality, and a strong moral code. Some extremely spiritual Muslims are called Sufis.

Katme told me that a sincere, practising Muslim becomes *angelic*, full of love and care to all humanity and to the animals and the environment.

> We all become one, and because we are all made by the same Creator, a spiritual bond develops between us. The outcome of this harmony is that everything in the universe

– animals, birds, plants etc. – all praise the Creator, and all are in a state of prayer, meditation and spirituality. Simplicity in life becomes apparent with responsibility in actions and a lively moral code. A Muslim is judged by their heart and actions and not by the appearance of religiosity. Islam, if followed completely, provides a structural framework that attempts to guide the whole of your life, even in medical ethics. I always seek guidance from Islam.

For Katme, peace throughout life is linked to spirituality. Along with formal prayers a Muslim is required to remember God in everything they do, by the use of special holy statements called *du'a*. For example, a Muslim thanks God and remembers him before and after eating, when he looks in the mirror, when he approaches his wife intimately. He says a particular prayer when starting out on a journey, as well as before sleeping and after waking. "In the Qur'an we learn that when you sleep you are in a state of death – your soul goes away," says Katme. "If God decides to keep you alive it returns when you wake, otherwise one will die in your sleep. This," he continues, "fits well with the concept of near-death experiences that many people report."

Also, when I go to my clinic, I remember God, so my medical practice should be based on honesty, sincerity, care and compassion and the moral divine code. I am conscious of God watching me and all my actions are recorded to be displayed and accounted for on the Day of Judgement. Obeying God and remembering the afterlife gives one a sense of tranquillity and a balance in spirituality.

The Qur'an also calls on Muslims to gain inspiration from every aspect or phenomenon of the universe.

One should look to the moon, stars and planets and see how beautifully they are designed fixed and functioning; to the animal kingdom, to see their fascinating organisation and behaviour; to the human embryo and the wonderful human body, which is like a temple, to worship and admire

God, the Creator and best Designer. One should look to plants, fruits and flowers to see how they develop and grow in a planned and organised way, to give you tasty, attractive, nutritious food, and to the rain and sky, to the day and night, to water formation, rain, river, sea. All these natural phenomena when observed and studied scientifically will lead you to believe more in God the only Designer, Creator and Sustainer of all creation.

Empathy

This concept of spirituality works well for people who believe in a deity, and want to use the human ability to look out of ourselves as a means of seeking a god and communicating with him. But people like Warnock believe that it is over-restrictive to confine our understanding of our spiritual nature in this way. For her a more secular understanding places it firmly in the realm of relationship. It is the human ability to relate at a deeply emotional level with other people – to empathise with their predicament.

An easy example, says Warnock, comes from Jane Austen's classic novel *Emma*. As in all of Austen's novels the plot centres on the question of who will marry whom, and why the coupling will occur. Will it be love, practicality or necessity? Heiress Emma Woodhouse seems pathologically obsessed with meddling in other people's affairs at the same time as attempting to scheme herself to her perfect match. About three-quarters of the way through the book Emma and friends go for an outing on Box Hill, a popular beauty spot 20 miles from the centre of London with a staggering view over the countryside. The group includes Mr Knightly, who Emma is just beginning to fall in love with, and Miss Bates, who Emma considers to be an inferior person who is welcomed along as a source of comic relief.

Sat overlooking the scene, they play a game. Everyone must say either one clever thing to Emma, two moderately clever things, or three dull things. When Miss Bates begins to chatter incessantly, Emma puts her down harshly. Mr Knightly is not impressed and scolds her for treating Miss Bates so rudely. "It's impossible not to be moved by the ghastliness of Emma's experience on Box Hill, her embarrassment, her misery because she is just beginning to fall in love with Knightly," says Warnock, hunching her shoulders in her own display of unease. "And when Knightly says, 'Emma, I should have not thought it possible ... ' – that is what I mean by spiritual."

This also brings us back to Warnock's belief in different ways of knowing truth. As you read the story you recognise the situation as true. It is a genuine reaction to an awkward situation. Although, of course, it is not true in the sense that it really occurred, but it tells you something that is true to human nature. The spiritual facet of our nature is the element of me as a human that is capable of relating to that story and recognising that the underlying elements are real. This challenges the purely scientific view of people like Richard Dawkins, who believe that there is only one way of knowing, that is a scientifically determined way, a way bound by measurable quantities and rationally solvable equations. All else, he says, is illusion. But what of toothache, bereavement, joy, bank overdrafts and debt? Your loan is as real as the car you bought with it. They all exist – they are true – but they have no physical existence. Science has its limits, and there is much more beyond.

Warnock continued,

> It says a lot about the commonality of human beings that they have an enormous amount in common at this spiritual level – the fact that everyone can be moved to tears by a performance of the *St John Passion*, or whatever it may be. It is a shared experience which shows that human beings have an enormous amount in common.

This is perhaps what Karl Marx referred to as our "species-being" – our recognition that we exist as an element that is bigger than ourselves as individuals. By using the term he was denying any division between individual and society: for him, human meant social. In Marx's view, we complete our individual and species character only by social interaction and over time.

Religion, wrote Karl Marx, is the sigh of the oppressed creature, the heart of a heartless world, the spirit of a spiritless situation.[14] In Marx's view, religion acts much like a pain-killing drug, something that lifts us away from the reality of our situation. Religion, he says in effect, is a false spirituality. His anxiety is that it will mask our awareness of the fact that we are bound together as a human collective; we will fail to discover ourselves as a species-being.[15]

I mentioned this empathetic use of the word "spiritual" to Rowan Williams. He was more than willing to recognise that people did use the term to describe a person who demonstrates a refined aesthetic awareness of poetry, music or the other arts, or perhaps one who is sensitive to the needs of other people. "This use has the advantage that it gives the appearance of being politically and religiously harmless and is therefore widely acceptable in our secularised culture," he commented, but went on to make it clear that he felt it was only a glimpse of the real thing.

I took this one step further when I met up with academic and author Kenan Malik. He has made a name for himself by his human-centred approach to philosophy, an approach that sees no room for anything as mystical as concepts of mind or spirit. All the same, he argues that a key element of humanity is our ability to have a subjective view of the world. "The consequence is that we can break out of the prison of our heads, to transform neurophysiological processes into subjective feelings and to understand the inner workings of others," he wrote in *Man, Beast and Zombie*, a book looking at human nature.[16] He develops an argument that human beings are more than their biology, that

---

[14] Reference 7 in Hay and Nye, *The Spirit of the Child*.

[15] Reference 8 in Hay and Nye, *The Spirit of the Child*.

[16] Malik K. (2000), *Man, Beast and Zombie*, Phoenix, p. 220.

their subjective assessment of the world around allows them to live in societies.

> Interactions between certain physical entities – human beings – give rise to non-physical entities that can have an impact on, and cause changes in, the physical world. Human beings are the only physical entities that can do this and they can do this not because they possess some magical powers, or because they live in some spiritual realm, but because, uniquely, they are not just physical but social beings as well.[17]

Many people would recognise the physical–non-physical interaction that Malik mentions; however, most would disagree with his conclusion, saying this is exactly where a spiritual link between people exists.

Duende

To say that a facet of what it is to be human is to be spiritual is an objective statement, a statement of fact. Associated subjective statements would then expand on this by talking about the nature of the spirituality, and here some people would want to add comments about good and bad, or even good and evil.

One place you can see this played out is in flamenco dance, and is summed up by the Spanish word duende, or in bullfighting where a critic commentating on a particular matador's performance might say that he had duende. This would not be saying that they are simply good at their art, but that they are from time to time possessed with a brilliance that enables them to perform out of their skin, and this performance stimulates the spectators as if the air was full of electricity. The term is almost exclusively used of performers whose heritage goes back to the Gypsies who migrated west from India around 1400. It is the sort of performance that speaks into people's hearts, that sends

[17] Malik *Man, Beast and Zombie*, p. 265.

tingles down the spine and causes audiences to rise en masse in a shout of praise and amazement. They recall the event as if time stood still.

There is a fascination about duende, but it is one that is tinged with horror. Both flamenco, and more so bullfighting, have death as central parts and when duende takes hold, the performers become enthralled by this darker side of nature. Consequently some people equate it with a form of demonic possession that the performers summon up in order to bring a spiritual quality to the occasion. This quality, when it arrives, transfixes not just the performer but the audience as well – everyone becomes involved. The master showman and legendary Italian violinist Niccolò Paganini (1782–1840) was said to play with the devil at his elbow. So great was the general public's fear of his demonic association that they delayed his burial in consecrated ground until five years after his death.

Morality

I began to wonder whether, if spirituality is so good at letting us think of others, and some would add seeking God, is it essential to moral thought? Is spirituality the component of our make-up that enables us to be morally aware?

"No," replied Warnock.

I think you could have a moral focus by considering that you are a member of society and considering that there are other people who depend on you and you on them – it wouldn't necessarily have any connection.

I would want to distinguish them. What starts people off on morality is the consciousness that we are all in the same boat. That what is nice and nasty for an individual is nice and nasty for everybody. This is a sort of common basis of morality; that if it is horrible for you to have your possessions stolen, or to be bullied, it is horrible for the other person as well. But you could have someone who understood that and lived by it, and changed his views because of it, but who nevertheless was actually totally unimaginative in other fields."

She went on to say, however, that "somebody who has no concept of the imaginative power or depth that I would comprehend under spiritual, would be lacking a dimension of being human. This is one of the things that human beings can amazingly experience."

Archbishop Rowan Williams held a similar line, but with a difference.

Obviously morality and ethics don't draw exclusively from the spiritual. There are any number of things that feed into an ethical code, which may include even things like common sense. But I would say there are some elements of moral life that can't survive indefinitely without a sense of the spiritual, and I mean by that something that works primarily in a negative way, rather than a positive way. If the spiritual includes looking a bit quizzically, a bit critically, at what I say about self, my needs, my priorities, my agenda; if the spiritual says, hold on, you know very little about yourself, and the truth about yourself is something in the hands of God rather than tucked in your mind, then your attitude to other people and your environment may be a great deal less aggressive, domineering, exploitative, or whatever. And so in that sense I think spirituality very fundamentally feeds into the moral. The spiritual aspect of humanity tells you to ask yourself awkward questions about what you think you want and what you think is your moral interest.

In his study of children's spirituality, David Hay found a similar connection.

At the level of practical politics the most important single finding of my research over the past 20 years is the very strong connection there appears to be between spiritual awareness and ethical behaviour ... Typically they say that the initial effect of their [spiritual] experience is to make them look beyond themselves. They have an increased desire to care for those close to them, to take issues of social justice more seriously and to be concerned about the total environment."[18]

The inverse of this is the privatisation of opinions and ethics. The call to let everyone form their own options and opinions reduces any chance of using our innate spirituality to change society because it is much harder to move from privately held personal beliefs to political agreement and subsequent legislation.

Education

In the closing pages of her book, *An Intelligent Person's Guide to Ethics*, Baroness Mary Warnock argues strongly for a revitalisation of spirituality within education. She claims that school should be a place where children learn the value of doing what is good, and where they can practise doing it, and even come to prefer it. Each pupil therefore will learn to do well and carry this into life. But in addition to this she hopes that educational institutions, like schools and universities, will be places where people can learn to exercise their spiritual as well as their physical and mental faculties.

There is sometimes an outcry against such talk, on the grounds that spiritual must mean religious ... But in truth any teaching is spiritual which opens a child's eyes to the position he as a human being occupies in the universe. In this sense, a lesson in palaeontology or geography, in biology, ecology or chemistry may be spiritual, insofar as the pupil may gradually learn where he stands in the history of the world, what he can try to find out what his responsibilities are.[19]

[18] Hay and Nye, *The Spirit of the Child*, p. 17.

[19] Warnock, *An Intelligent Person's Guide to Ethics*, p. 179.

"I think we neglect the aesthetic, the spiritual, that which is incapable of being expressed in words, we neglect all that very much at our peril. Because we are in danger of giving people an almost totally mechanistic view of themselves, they are not prepared to believe in the kind of imaginative excitement or think it worth pursuing," she told me, adding that in her view, our current emphasis on having an educational curriculum packed with functional subjects is a denial of spirituality. Telling someone that the whole reason of going to university is to get a well-paid job again absolutely denies the spiritual element of people. This she feels is especially sad, as people in adolescence are capable of being particularly imaginative – they are specifically equipped to exercise their spiritual faculty.

David Hay's research found that children were particularly capable of using the spiritual component of their being, often expressing it in young children's screams of excitement or quiet wide-eyed wonder as they explore the world they are part of. But, he also says, it is easy to miss.

> One needs to enquire carefully about and attend to each child's personal style if one is to hear their spirituality at all. At a theoretical level this implies that we cannot neatly distinguish the spiritual aspects from the psychological features of a child's life. [But if you listen,] children can provide their own evidence of what their spirituality is like.[20]

If you are ever in doubt about a child's sense of imagination, just give them a box to play with and see how it mutates from a box to a boat, and from a castle to a car. And learning to use your imagination will exercise your ability to use your spiritual faculties.

Passing through teenage years, however, is all too often destructive to this process.

> Children emerge from infancy with a simplicity that is richly open to experience, only to close off their awareness as

[20] Hay and Nye, *The Spirit of the Child*, p. 122.

they become street-wise. To be open is to be vulnerable. Its contrary is to know the score, to know how to look after yourself in a hostile environment.[21]

Hay is distressed that school science is now packaged in plain facts, and so any sense of awe and wonder is squeezed out. It is a thought echoed by James Lovelock, inventor of Gaia theory, who reminisced, when I met him, of his school chemistry.

> There were beautiful refractive liquids you could hold and see the sunlight go into a rainbow of colours. There were amazing smells of every description. There were bangs. But it has all been wiped out and now chemistry is taught by dull computer simulations on computer screens and by dull uninspired stuff – no wonder nobody goes into it. You wouldn't think of chemistry as spiritual normally, but I did.

Rowan Williams said,

> One of the things that happens around puberty is that we become acutely aware of our physical selves and therefore of our emotional selves in a quite new way. That acute self-awareness is one of the things that, by making us a bit suspicious of ourselves, even puzzled by ourselves, can lead us to think that earlier childhood experience is somehow just that, a childhood experience. We have grown out of it. We have become a different sort of self around puberty ... That is the point at which the young person, the pubescent teenager, most readily gets bored, disillusioned and uncomprehending about the language of the spiritual.

And the archbishop is one of the first to recognise that as people throw away an expectation of experiencing the spiritual facet in their life, they throw away religion.

[21] Hay and Nye, *The Spirit of the Child*, p. 122.

For Warnock, this is the clue to why society is throwing away any sense of morality; she told me:

> People think that if they throw away religion they can throw away morality too. I think one has got to find a way of arguing for the essential nature, essential part of being human, that is being moral quite separately from religion although of course there are lots of religious people who derive their morality from religion. But the point is there are lots of people who need to be taught to be moral, to think morally, who reject religion, and they shouldn't be allowed to throw out the baby with the bath water.

"In most state education systems spiritual education as a cross-curricular element is still a distant dream. For it to become a practical reality there will have to be a radical change in educational culture," says Hay. "Genuine social integration arises from another source: a widespread awareness that relational consciousness is the bedrock of a free and humane society. In such a society the primary task of education is the nurture of the spirit of the child."[22]

At work

Go into any bookshop and have a look at the shelf titles. It will not take you long to find a spirituality zone of one form or another. Business spirituality is particularly popular at the moment. Around the UK there are about 500 courses in management studies, and over 200 include something on spirituality.

Archbishop Rowan Williams was deeply sceptical. "My scepticism about it comes from a great suspicion of using spirituality simply as a means of doing things better," he told me.

> Spirituality for the business man or the cook? If it simply means developing a nice sort of clear-headed, self-possessed approach so that you can make more money or cook nicer

[22] Hay and Nye, *The Spirit of the Child*, p. 175

meals, well that is all very well. But spirituality makes you ask awkward questions about yourself, and it doesn't sit very easily with quite that model. Now, the defensible side is that there is a spiritual dimension to everything, it is what Buddhists call mindfulness, having a clear-headed awareness of what is actually going on, what you are actually doing. Not being tense, anxious, distracted with your mind on 35 things, trying to do three things at once. Yes, part of the spiritual is deep breathing, it is saying, "Yes I am here, and this is what I have got to do, and my mind is present in my body." Now that is OK, that is spiritual, but it is only a bit of it.

Again, a historical example. St Teresa of Avila, who is one of the classic cases of a mystical life history. [Born in 1515, Teresa had entered a Carmelite convent at the age of 18, and by 1555 she had gained notoriety for extraordinarily intense spiritual experiences.] She describes in her writings about her own experience, how at the beginning you are knocked sideways and funny things happen to your body and your ideas, you get disorientated and you get close to a breakdown. It happened to her. And for quite a lot of her life, she says, that is what it was like. She had an unusually long period of very disruptive and painful and rather dramatic spiritual experience – visions. She even talks about a levitation. She felt it was a great burden and at some stage wanted to be dead – there was an overload thing – it was too much to carry.

If this is a true expression of radical spirituality it is hardly what you need if you want to increase your business performance or your chances of making the perfect soufflé.

The breakthrough for Teresa came in the 1560s when she found that she was able to relax where she was – when she felt perfectly at peace without any sense of anxiety that she should be somewhere else or doing something else. This profound sense of satisfaction, she says, is one of the characteristics of the deepest states of union with God. Most of us live remembering good times in the past and longing for a future when life is more comfortable, but forgetting that the only place we live is now. "Union with God is not flitting off in mystical rapture, it is being in unshakable consciousness, here and now; knowing that there is no gulf between you and God," commented Williams. Intriguingly, he pointed out, Teresa says that once she had discovered how to be fully comfortable with herself, to reach this level of spiritual fulfilment, she was a better administrator. So maybe there is a hope for business spirituality, but if Teresa's example has anything to show us, it is going to demand a lot more than just getting good dynamics within an office.

Possibly this is where Eastern forms of organisation such as feng shui come in, as they help a person make sensible use of their working environment, and create an atmosphere in which our spiritual natures can operate well. As such, feng shui is not spiritual itself, but enables our spiritual existence. "Look at the 1930s studies in Chicago about how the colour of the environment changes people's mood," says Edward Bailey, adding that many churches have recognised this in the way that they change the colour of the material over altar tables, lecterns and pulpits so that the colour matches each season's mood, and choosing the colour of your clothes is one way of telling people about your inner self.

But how far can we stretch the concept? If we can talk about spiritual assistance in the workplace, can we start to think in terms of spirituality in other realms of life – in non-human animals, in trees, in a sunset over a calm sea?

To start off with, says Bailey, people have often dressed and erected buildings that are consonant with their environment. "Look at Masai dress and the way it echoes their environment – it may not be articulate, but it is intuitive."

Bailey also has no problem with the idea that you could use the term "spiritual" for non-human animals, so long that you don't think it implies all that the term means for human beings.

I am sure we can talk about spirituality of animals, or even trees, but obviously it is a different spirituality. After all the spirituality of a one-year-old baby is different to that of a seven-year-old child, which again is different to an adult. They are all alive – so in traditional theological terms God is within them bringing them to life. It doesn't take us far, but it does give us a different attitude to our environment.

He was keen to stress that this in no way suggested an equality between humans and other animals, as is argued by people like Australian philosopher Peter Singer.[23] "I can't stop them saying it, but they could say the moon is cheese if they like. They are free to say it, but I can't for one minute go along that line," said Bailey. "There is all the difference in the world between saying that man is an animal and saying that man is nothing but an animal."

The Archbishop considered these ideas when I put them to him. "There are very few mystical autobiographies from hamsters and that creates a bit of an evidential problem here," he said with a mischievous glint in his eye.

You see, I don't know how we could know really, because we have absolutely no means of knowing what is going on inside the subjectivity of an animal. People talk sometimes about the behaviour of some animals – dogs often – showing loyalty, affection, heroism and so on, and I don't rule that out. Who knows? But there is no way of systematising that. What we do know is that one way of characterising us as human beings, as far as we can see, is this capacity for looking at ourselves in a certain way and opening ourselves up to the divine.

I asked about the biblical language of creation worshipping God.[24] Doesn't this make animals, plants, even the very rocks and seas,

---

[23] For an example of his thinking, read Singer P. (1994) *Rethinking Life and Death*, Oxford University Press.

[24] For example, Psalm 98:7–9; Romans 8:19–22.

spiritual in some way? "Good point – I quite like that language," he replied.

God made the world, in Christian theology, so that the world will find joy and fulfilment in relation to God. We know about ourselves as Christians – that is where the fulfilment lies. Now what is the fulfilment of a dog, a hamster, an amoeba, a rhododendron bush or Mount Everest? I don't know. But to talk about creation worshipping is to say that the very existence of those realities in some sense reflects back to God what God has given, and by analogy with our own experience we talk about joy or praise. And there is a very strong Christian tradition, especially in the Christian East, which says that the task of humanity within that picture is to be the kind of choirmaster, to give voice to the way that creation reflects back the beauty or the joy of God. That is why we have a responsibility to the rest of the created world to treat it with respect, to use it in ways that don't somehow insult the divine Creator. So to use it acquisitively and exploitatively and violently is somehow to intrude on the joy and wisdom and the balance that God has put there. And to use it with reverence and to celebrate it as well, artistically in word and image, that's part of our relation to the created world. That is possibly where you link up with Baroness Warnock and the imagination, that there is a way of relating to the environment around you which whether you know it or not is somehow spiritual and celebratory.

There is a sense in which you could say that the spiritual evolves as human self-awareness evolves, as language and the capacity to picture yourself evolves. Just as the individual, so in the species. I guess. Which is why there is no way of ruling out other kinds of animate life developing in that direction.

All the same, he was deeply uncertain about any idea that this could come via a neo-Darwinian process of survival value.

Likewise Katme looked into his Islamic religion and concluded that some form of spiritual intuition could not be ruled out in animals and trees. "The Prophet Muhammad, peace be upon him, used to stand and preach to people from behind a tree. Then they got a mosque and he moved in there – the trunk cried for the loss of the prophet," he told me. On another occasion a camel started to cry when he met the prophet, causing this holy man to deduce that the owner was cruel.

Twenty-first-century culture seems to be at a curious place. On the one hand there is a loss of respect and trust in the established forms of religion, but on the other hand there is an increased interest in spiritual issues. In science and medicine you can see the paradox played out. Research exercises, such as the human genome project, have bred a new set of reductionist scientists who believe that if it isn't physical, it doesn't exist, while at the same time there is a burgeoning use of so-called alternative medicines, many of which base their claims for success on a spiritual understanding of our existence.

The people quoted within this chapter have highly differing views of how the spiritual side of a person's nature is best employed, but one point of agreement between Williams, Warnock, Katme and Bailey was that the decreasing interest in religion is leading people to forget or misunderstand their own spirituality. They parallel this change with the breakdown currently experienced in Western society, where we increasing live as individuals and attempt to ignore our neighbours.

If spirituality is about developing relationships, either kinship relationships or relationships with the divine, as well as giving us an ability to appreciate our commonality, then not exercising this facet of our existence will lead to an impoverished life, because at root it will deny that we are spiritual beings.

CHAPTER EIGHT

a sexual being

It drips from advertising hoardings and adorns the front of glossy magazines. It drives some people to perform madcap ordeals to prove their passion and others to commit murder. It is both talked about and taboo at the same time. "It" of course is sex.

"Sex is the mysticism of materialism and the only possible religion in a materialist society," said the late journalist and sage Malcolm Muggeridge. When I met him at his home in Kent shortly before he died in 1990, he explained that he had given up watching TV news because he claimed its producers were more interested in the sexual attractiveness of the presenters than concerned about the content. In fact he had gone further and had completely given up watching television for similar reasons.[1] He was, however, aware of the paradox in that he frequently appeared on television. In one interview he said, "On television I feel like a man playing a piano in a brothel; every now and again he solaces himself by playing 'Abide with me' in the hope of edifying both the clients and the inmates."[2] And his scorn for the way that sexuality was misused reached outside the box.

> In America today the religion of sex has reached its highest point of development and attracted the largest numbers of adherents. The devout dedicate their lives to sex, from cradle to grave and even beyond. The sacred tests, from *Kama Sutra* to *Fanny Hill*, *Chatterly*, the *Cancers* and *Candy*, edify the faithful ... Eat this, wear this, anoint yourself with this, in remembrance of sex ... it is of sex that the cherubim and seraphim continually do sing. In the beginning was the Flesh, and the Flesh became Word.[3]

[1] "The only thing to do is to follow me and have your aerials removed; a very simple operation, it doesn't hurt in any way, and you'll feel infinitely better for it." Muggeridge M. (1967), *Ancient and Modern*, London: British Broadcasting Corporation, p. 174.

[2] Muggeridge M. in a television interview, September 1972. Quoted in *People on People: The Oxford Dictionary of Biographical Quotations* (2001), Oxford University Press.

[3] Muggeridge, *Ancient and Modern*, p. 184.

Sex, said Muggeridge, sells everything from deodorants to bulldozers. Why? Well, the bottom line is that sex has this amazing power over us because we are sexual beings. You may argue that you are homosexual, bisexual, heterosexual or transsexual, but I guess not many say they are asexual. And even they are still using the concept of sexuality as a means of understanding themselves, albeit in a negative manner.

As we saw in Chapter 7, sexuality has a strange parallel with spirituality. Both are most certainly part of our make-up, but the way that they operate and are applied to life has become highly contentious. Some people hold that there is only one way to view sexuality, just as there is only one correct use of their spirituality. Others say that they are aspects of our being that should be left to the individual to decide how to respond, rather than have some societal expectation forced on them. With sexuality some claim there is only one way to use it properly, normally advocating that sexual activity is confined to married heterosexual relationships, while others maintain that it is down to personal choice. But no one seriously claims that humans are not sexual beings.

Clear terms

The fear of pressure to conform to preconceived societal norms, leads to one of the greatest problems that anyone who tries to discuss this area faces. In *The Blank Slate*, Steven Pinker lists gender in the "Hot Buttons" section of his book.[4] And he is right to do so. When religion was the key issue that people debated, you could get burnt at the stake for stepping out of line. Now sexual behaviour has replaced it as the area in which it is most dangerous to question presumed societal norms. Anyone who dares ask questions of human sexuality lays themself open to, at the very least, a metaphorical roasting.

The area has become so politically dangerous that some academics argue you should not research into aspects of it. One

[4] Pinker S. (2002), *The Blank Slate*, Penguin, Chapter 18.

of the areas where this crops up prominently is the discussion of the scientific basis of sexual orientation, and in particular the reasons why some people are homosexual. "This is an area, *par excellence*, where scientific objectivity has little chance of survival," says John Bancroft, director of the Kinsey Institute for research in sex, gender and reproduction, in Bloomington, Indiana.[5] In effect he says that everyone brings so many preconceived ideas to their work that none of the results can really be trusted. A cause and a consequence of this is that, in comparison with many other areas of investigation, the scientific study of sexual orientation is, at best, still in its infancy.[6] Sadly, that situation is unlikely to change fast. The problem is clearly displayed in a 1996 review of the current state of biomedical research on homosexuality. This concluded that so far the causes of homosexuality are unknown; that sexual orientation is likely to be influenced by both biological and social features, and that the area could be studied. So far so good. The review then argued that research into the causes of homosexuality would be *unethical* and *should not* occur.[7]

This has a subsequent problem. If you can't study why some people believe they are homosexual, you severely restrict the ability to look at why others are certain they are heterosexual. In fact, by placing an embargo on this politically awkward area, you inhibit any thorough examination of sexuality.

According to Pinker,

> [the] politics of gender is a major reason that the application of evolution, genetics and neuroscience to the human mind is bitterly resisted in modern intellectual life. But unlike other human divisions such as race or ethnicity, where biological differences are minor at most and scientifically

[5] Bancroft J. (1994), "Homosexual orientation: the search for a biological basis", *British Journal of Psychiatry*, 164, pp. 437–440.

[6] Byne W. and Stein E. (1997), "Ethical implications of scientific research on the causes of sexual orientation" *Health Care Analysis*, 5, pp. 136–148.

[7] Schuklenk U. and Ristow M. J. (1996), "The ethics of research into the cause(s) of homosexuality", *Journal of Homosexuality*, 31, pp. 5–30.

uninteresting, gender cannot possibly be ignored in the science of human beings.[8]

To talk about the issue clearly, there is a need to sort out a few terms that are commonly used. Looking at them not only helps us understand the comments made by each contributor to a debate, but also gives us an insight into the minds of the people using the terms. The first is "gender assignment". The vast majority of people are assigned to one gender or the other at birth as the midwife and parents make a visual inspection of external genitalia. On rare occasions, this external signature of identity is malformed and the decision is not straightforward. Medical science now allows doctors to analyse cells from the baby, normally collected by taking a small blood sample, and look to see which sex chromosomes are on board. The baby then is assigned to one or other sex. Even with this technology there are cases where doctors still cannot make a clear-cut decision, and there is a growing feeling that these children should be allowed to decide which sex they feel most at home with later in life – as we will see later, these situations occur, but they are extremely rare.

The next term to consider is "gender role". This becomes more contentious as it refers to the behaviours that are more associated with men (male gender roles and behaviours) or women (female gender roles and behaviours). This is the area of debate most involved in the nature-nurture debate. Are differences in role simple reflections of the way that women and men are made, or are they impositions conjured up by society – normally seen as a male-dominated society that seeks to keep women "in their place"?

How people see themselves is referred to as their "gender identity". It is the total perception of an individual about their own gender. This will include whether they feel they are a boy or girl, man or woman, as well as some form of assessment of how

[8] Pinker, *The Blank Slate*, p. 340.

well they feel they conform to the social norms of masculinity and femininity.

Finally there is the idea of "gender attribution". This is what we do to others as we meet them in the street or at work. We use a variety of clues to decide whether the person is a man or a woman. These include mannerisms that can be socially as well as physically influenced; the way a person walks and moves, clothing, use of make-up, the style of jewellery, and at times even the type of job that the person holds.

For the majority of people the category used at each point will be identical. For example, I was assigned male at birth, feel that I identify with the role that society tends to expect of a male in twenty-first-century Western Europe and as far as I am aware people have no difficulty recognising me as male when we meet. That said, it is still a useful exercise to note that there are different practical ways that we can assess sexuality.

Vive la différence

Another landmine-strewn area to wander through is the territory that discusses whether there is any inherent difference between men and women, and if so how these differences came into being. The average person walking down the street seems to have little problem coming to terms with the idea that men and women differ, though they may argue about the exact nature of the difference, but in academic circles the debate has become highly contentious.

"Why are people so afraid of the idea that the minds of men and women are not identical in every respect? Would we really be better off if everyone were like Pat, the androgynous nerd from Saturday Night Live?" asks Pinker, before going straight on to answer his own questions. "The fear, of course, is that different implies unequal – that if the sexes differed in any way, then men would have to be better, or more dominant, or have all the fun."[9]

[9] Pinker, *The Blank Slate*, p. 343.

There is nothing new with the underlying issue in this dispute. Value and power have all too often gone hand in hand. Physical power means that you can dominate, you can extort things from others and you can elevate your personal lifestyle at the expense of your compatriots. Much of recorded history is a tale of power and persuasion. Looking to sexual politics, the male's ability to hit harder, to run faster and to avoid prolonged servitude towards his offspring has led to that half of the human species seizing power.

There is no point denying that differences exist, or trying to bend data so that they give the impression that there are no gender differences. While there is a good degree of variation that inevitably creates a measure of overlap, men tend to be taller than women. They can also reach more physical extremes of power. Just look at the leader boards at an Olympic Games. All of the men's times are shorter that the women's, the heights greater, the lengths longer (see Table 8.1). The same is reflected in world records. The men's world record for 100m is held by American Tim Montgomery at 9.78 seconds and the women's by Florence Griffith-Joyner at 10.49 seconds. The marathon record for men is 2 hr, 5 min 38 sec, and for women is 2 hr 15 min 25 sec. It doesn't mean that the men are any more deserving of their medals, nor that they have worked any harder than their female counterparts; just that the relatively small differences in the way men and women are built gives men a distinct advantage.

It's not all one-sided once you move away from brute power. Males have a greater tendency to have dyslexia, learning difficulty and attention deficits, particularly when they are younger. Women have better verbal skills. Men are less sociable and more aggressive and more prone to downright violence. Even after a distinct increase in female involvement in murder, men are ten times more likely to kill someone than are women.[10] In contrast, women are more likely to show aggression towards

[10] Tye L. (1998), "Girls appear to be closing the aggression gap with boys", *Boston Globe*, 26 March.

Table 1 World records as of April 2003

Event	Men's world record	Women's world record
100m	9.78 sec Tim Montgomery Paris; 14 September 2002	10.49 sec Florence Griffith-Joyner Indianapolis, IN; 16 July 1988
Marathon	2 hr 5 min 38 sec Khalid Khannouchi London; 14 April 2002	2 hr 15m 25 sec Paula Radcliffe London; 13 April 2003
Pole vault	6.14m Sergey Bubka Sestriere; 31 July 1994	4.81m Stacy Draglia Palo Alto, CA; 9 June 2001
Javelin throw	98.48m Jan Zelezný Jena; 25 May 1996	71.54m Osleidys Menéndez Réthymno; 1 July, 2001
Hammer throw	86.74m Yuriy Sedykh Stuttgart; 30 August 1986	76.07m Mihaela Melinte Rüdlingen; 29 August 1999

Data from http://www.iaaf.org

other people through ostracism or alienation than physical violence. Being prone to feel less secure, men compete harder than women for social status, and in the West they link their sense of self-esteem to salary, wealth and possessions – a feature that gives men in some cultures the tendency to view their wives as yet another form of possession.

In biological terms there are the obvious differences in sexual organs, but some anatomists now take this further, claiming that there are small, but genuine, differences in the structure of men's and women's brains. The differences exist, and books like John Gray's *Men Are from Mars, Women Are from Venus*,[11] make an

---

[11] Gray J. (2002), *Men Are from Mars, Women Are from Venus: How to Get What You Want in Your Relationships*, HarperCollins.

enjoyable read as well as performing the role of a self-help guide for people trying to relate to members of the opposite sex.

The question then is: Where do these differences come from? What is it that shapes this important aspect of the person I call me? If we go back to our school biology classes, then life is simple. In one of our very first lessons we discover that almost all animals come in two types. There are males and females. Moving on in our academic careers we learn that the biological difference is equally simple. At the heart of each cell in plants and animals is a nucleus containing pairs of chromosomes. These chromosomes carry code that enables our bodies to develop and operate smoothly. One pair is particularly involved in determining the overall sex of the individual – the "sex chromosomes". Women have two "X" chromosomes, and men have one "X" and one "Y". The current thinking is that there are a few genes present on the Y chromosome that direct an embryo to develop as a male. If these signals are absent, or for some reason they are sent but not received by the cells, the embryo will grow as a female. You can see this in children with Turner's syndrome, who have only a single X chromosome and no Y, and develop female bodies.

In the vast majority of cases this biological component of our being is reflected at birth by the presence of male or female genitalia, and can be clearly recorded on an individual's birth certificate. Equally, in the vast majority of cases, this physical appearance is also related to our growing perception of who we are. "I'm a boy," says my young son as he defends his argument for having only boys to his birthday party. "We hate girls." And for the most part, girls are equally reluctant to invite boys. It is part of the emerging awareness of identity and difference.

Bedroom and boardroom power

One aspect of the problem is that as soon as you mention sexuality many people confine their thinking and responses to issues that relate primarily to physical sexual acts. This would be a similar class of mistake to assuming that a review of spirituality is totally confined to a discussion of religion. Sexual activity is obviously going to be part of the outworking of a person's individual understanding of their sexuality, just as religious affiliations and experience are often part of the outworking of a person's spirituality but, in both cases, the obvious manifestations of these aspects of our being are only a part of the real thing.

That said, for all but those who have decided for whatever reason to live a celibate life, sexual attraction and intercourse is an important outworking of their sexuality.

The 1970s are often described as an age of sexual revolution, a time when society started to throw off the straitjackets of taboos and restrictions. People started to experiment with a new sense of "freedom". Sex was permissible before marriage, and became more acceptable outside marriage. Politicians started to dismantle some of the laws dictating that only heterosexual activity was permissible. Society dismantled its prohibition against sex outside marriage.

These changes coincided with a growing understanding of the ways that hormones within the body enable a woman to become pregnant and provide an environment in which the developing baby can grow. Once this was known, it was only a matter of time before someone would work out, or stumble upon, a chemical that could disrupt the process. The "someone" turned out to be Carl Djerassi, and the chemical was 19-nor-17alpha-ethynyltestosterone, norethindrone for short, a steroid that he derived by chemically altering a molecule extracted from the juice of an inedible, hairy, Mexican yam.[12] He discovered it on 15 October 1951, and the end result a few years later was "the pill". Djerassi insists that the pill did not create the sexual revolution, but just fitted in with the times. Without the political desire to give women more control over their reproductive lives, this would have remained a drug of limited interest.

---

[12] Carl Djerassi tells the story of this discovery in his book, *This Man's Pill: Reflections on the Fiftieth Birthday of the Pill* (2001), Oxford University Press.

It would, however, be wrong to give the impression that prior to this turning-point in history, human beings were sexually "well behaved". Despite the best pretences of our Victorian forebears, sexuality had never been kept in the neat little box that even they hoped to achieve. In ancient Greece if you wanted a sexual encounter you went to the temple where religious prostitutes, male and female, were ready and waiting to serve you. Throughout history, rich young males have always seen poor attractive females as fair game for a brief fling. While on rare occasions the women concerned may have gained some enjoyment, the typical picture was that the male expressed the physical side of his sexuality and then walked away from the consequences.

At the same time, there is no reason to believe that homosexuality or other variations in sexual behaviour are new. Once again, temple prostitutes of ancient Greece came in all "shapes and sizes", and history records plenty of people who rejected mainstream notions of sexuality. What is new is the post-modern relaxation of any sense of community-agreed values, which allows individuals greater freedom to come out of hiding and live out their own notions of sexuality.

Rumbling through the whole of the twentieth century has been the feminist movement, which among other things campaigned for sexuality to step out of the bedroom into the boardroom. The pill meant that women were no longer confined to child-rearing and home-keeping. At the same time there was a growing realisation that girls should have equal access to education and that women should be given equal opportunities to succeed in the workplace.

In reality there is still a long way to go before that is realised in practice. This is in part because we are still not taking sexuality seriously. We allow, sometimes reluctantly, women into offices, clubs and boardrooms, but preferably if they will pretend to be men. Workplaces and work patterns have been shaped by centuries of male domination. If women are going to be allowed in as women, then the systems will have to change, and that will be more than just making sure there are more women's toilets in the building, or allowing the odd pot of flowers to appear around the place.

At the moment, people are being treated as if their sexuality does not influence them or the way they operate. This is despite the fact that a person's sexuality is important to the way he or she thinks, works and relates to others. In calling for equality, we are in danger of demanding monotonous uniformity. It is as if we are pretending our workplaces are zones where we can airbrush a person's sexuality out of their character; as if we want people to become sexually neuter as they enter the office.

Nature nurture

If sexual politics has clouded any discussion about the differing roles and mentalities of biological males and females, then party politics has done the same for the debate about underlying causes of these distinctions. The question is whether our gender assignment and gender identity are biologically imprinted features of our existence, or socially carved aspects of our lives.

It is interesting to see what has happened in the political realms of social discussion. Normally those whose politics lean to the "left" argue that people are shaped by society to perform the way they do. Change society and you will change people. On the other hand, those on the "right" are prone to say that intelligence runs in families, as does musicality and criminality. You are born the way you are, and it is society's job to employ your skills or restrain your disruptive tendencies. With sex the two swap sides. The left wants to argue that you are born with your particular sexuality, so no one has the right to question what you do. The right says that there are norms that should be adhered to and that – given the correct social environment, help and counselling – a person's sexuality can be reassigned so that they fit in with the majority.

As so often happens, this fight becomes personalised. One of the most famous examples of this is the so-called John/Jane case; fictitious names used to hide the patient's real identity, but a medical history found in many a textbook. In 1965 identical twins Bruce and Brian (their real names) were born to young working-class parents in Winnipeg, Canada. Bruce was the more active twin, Brian more peaceful and less rowdy. On 27 April 1966, when they were eight months old, the twins went back in to St Boniface Hospital, because the foreskins over their penises were too tight and they were both crying each time they urinated. They needed a circumcision – a simple operation. But for Bruce things went badly wrong. An electro-cautery was used, which seals blood vessels as it cuts, but during the procedure the boy lost the majority of his penis.

His parents were understandably distraught and had no idea what to do next. For advice they turned to sex researcher John Money who worked at the Johns Hopkins Medical Center in Baltimore, USA. His view was simple. Sexuality, he claimed, was conditioned, and not an innate part of a person's identity. If they made the child physically female, and gave him the right nurture and the correct hormones, they could turn Bruce into a happy, healthy (albeit infertile) female. He advised them to have the boy totally castrated and get surgeons to build an artificial vagina. So, at 22 months of age, Bruce was castrated and started a new life as Brenda. His[13] parents then continued to raise him as a girl without telling him his history.

In a review of the case written 30 years later, Milton Diamond, from the department of Anatomy and Reproductive Biology at the University of Hawaii at Manoa, and psychiatrist Keith Sigmundson pointed out that this advice was the common currency of medical textbooks:

> The decision as to how to proceed typically follows this contemporary advice: "The decision to raise the child as a male centers around the potential for the phallus to

[13] You may feel uneasy about my use of the male pronoun throughout the rest of this story. With Brenda being brought up as a girl, there would be grounds for using "she" and "her", but as his story makes clear, Brenda never really saw himself as a girl, so I think "he" and "him" are more appropriate.

function adequately in late sexual relation" and "Because it is simpler to construct a vagina than a satisfactory penis, only the infant with a phallus of adequate size should be considered for a male gender assignment." These management proposals depend on a theory that says it is easier to make a good vagina than a good penis and because the identity of the child will reflect upbringing and the absence of an adequate penis would be psychosexually devastating, fashion the perineum into a normal looking vulva and vagina, and raise the individual as a girl.

This management philosophy is passed on as two beliefs held strongly enough by pediatricians and other physicians to be considered postulates [i.e. taken for granted]: (1) individuals are psychosexually neutral at birth and (2) healthy psychosexual development is dependent on the appearance of the genitals. [14]

Over the years Money made many public statements about how well "this little girl" was getting on with her given gender identity. He used this case to build an apparently compelling argument that with the correct management doctors could successfully reassign a baby's sex. Money claimed that Brenda was sailing contentedly through childhood as a genuine girl. "There isn't any question which one is the boy and which one the girl," Money told Baltimore's *News American* newspaper. "It's just plain obvious."[15] And here was living proof that a person's gender identity and role are ultimately determined by environment. "The girl's subsequent history proves how well all three of them (parents and child) succeeded in adjusting to that decision."[16]

[14] Diamond M. and Sigmundson H. K. (1997), "Sex reassignment at birth: long-term review and clinical implications", *Archives of Pediatric Adolescent Medicine*, 151(3), pp. 298–304.

[15] Quoted in Colapinto J. (2000), *As Nature Made Him*, New York: HarperCollins, p. 78.

[16] Money J. and Tucker P. (1975), *Sexual Signatures: On Being a Man or Woman*, Boston MA: Little Brown & Co. Inc, pp. 95–98.

The dogma gained such widespread support that it even got into articles in publications such as *Time* magazine.

> This dramatic case provides strong support for a major contention of women's liberationists: that conventional patterns of masculine and feminine behavior can be altered. It also casts doubt on the theory that major sex differences, psychological as well as anatomical, are immutably set by the genes at conception.[17]

As a consequence of this persuasive presentation many other doctors went ahead and made similar recommendations, meaning that other boys were surgically reassigned.

By the time Brenda got to school the reality was somewhat different. Rather than being happy and healthy, he was an awkward child who did not engage in girlish activities and was mercilessly teased by schoolmates for his gunslinger stride and lack of interest in boys. When Brenda questioned it, the doctors simply replied that he was a tomboy. No one mentioned the truth.

Brenda fiercely fought against his gender identity as a girl. In Canadian journalist John Colapinto's book *As Nature Made Him*,[18] Brenda's mother tells of how she recalls him tearing off frilly dresses and preferring guns to dolls. He even insisted on urinating standing up; an action that got him into trouble at school. As he entered his teens Brenda's shoulders, neck and biceps became more muscular and his voice showed signs of beginning to break.

> There were little things from early on. I began to see how different I felt and was, from what I was supposed to be ... I thought I was a freak ... I looked at myself and said I don't like this type of clothing, I don't like the types of toys I was always being given. I like hanging around with guys and

---

[17] *Time*, 8 January 1973, quoted in Colapinto, *As Nature Made Him*, p. 69.

[18] Colapinto, *As Nature Made Him*.

climbing trees and stuff like that and girls don't like any of that stuff.[19]

At 14, Brenda was so miserable that he decided either to live life as a male or to end it. After being taunted at school for his odd behaviour Brenda got into a fight and was expelled from school. He was appalled at the idea of taking hormones so that he would sprout breasts and develop wide hips. At this point his father finally came clean and told the truth. "All of a sudden everything clicked. For the first time things made sense and I understood who and what I was," he commented.[20]

Brenda underwent a new set of operations to remove his artificially induced breasts and assumed a male identity. By the age of 16 he had been returned to as close to an anatomical male as possible. By 25 he had taken on a third name, David, and was married. "It was ignorant of them to think that I was no longer a male because my penis was burned off," he said. "A woman who loses her breasts to cancer doesn't (become) any less of a woman."

Colapinto wrote David's biography after seeing an article in the *New York Times* in 1997 which had been triggered by Diamond and Sigmundson's academic publication.[21] Diamond had been in a good position to cast judgement, as he was part of the research team at the University of Kansas which in 1959 identified that testosterone has a masculinising effect on guinea pigs while they are still in their mother's womb. Similarly Sigmundson had good reason to give an authoritative account as he had been Brenda/David's psychiatrist in his hometown of Winnipeg.

The medical community then faced the awful reality that hundreds if not thousands of individuals may have been sexually

---

[19] Quoted in Diamond and Sigmundson, "Sex reassignment".

[20] Quoted in Diamond and Sigmundson, "Sex reassignment".

[21] Diamond and Sigmundson, "Sex reassignment". He also published a follow-up article responding to the numerous letters he received once the case was in the public domain: Diamond M. (1997) "Sexual identity and sexual orientation in children with traumatized or ambiguous genitalia", *Journal of Sex Research*, 34(2) pp. 199–211.

reassigned into misery. The web site for the Intersex Society of North America (ISNA) reveals further, albeit anecdotal, evidence of problems with sex reassignment.[22] "Here, reassigned females report overwhelming confusion during childhood and adolescence, and either bisexuality, homosexuality, or avoidance of sexual activity. All are angry at the surgeries they endured without their consent or understanding," says US journalist, Ricki Lewis. "In contrast, sex reassignment surgery works well for transexuals, whose gender identity is at odds with their chromosomal and gonadal sex, and who request the surgery."[23]

"The medical treatment of [individuals who have no clear sex at birth or have suffered a] traumatic loss of the penis has been a 40-plus-year, poorly run, unethical experiment," says assistant professor of science and technology studies at Michigan State University in East Lansing, Alice Dreger. [24]

In the last couple of decades the pendulum has swung away from notions that personal identity is totally shaped by our upbringing, and biological factors are considered of increasing, if not of paramount, importance.

> Things are not looking good for the theory that boys and girls are born identical except for their genitalia, with all other differences coming from the way society treats them. If that were true, it would be an amazing coincidence that in every society the coin flip that assigns each sex to one set of roles would land the same way ... The two key predictions of the social construction theory – that boys treated as girls will grow up with girl's minds, and that differences between boys and girls can be traced to differences in how their parents treat them – have gone down in flames.[25]

[22] http://www.isna.org

[23] www.the-scientist.com/yr2000/jul/lewis_p6_000710.html

[24] www.the-scientist.com/yr2000/jul/lewis_p6_000710.html

[25] Pinker, *The Blank Slate*, p. 350.

There is increasing evidence then, that a large part of your sexuality is innate – you are born with it. It is part of your being, and it will affect your decisions. "The question of why more women don't choose careers in engineering has a rather obvious answer: Because they don't want to. Wherever you go, you will find females far less likely than males to see what is so fascinating about ohms, carburettors, or quarks. Reinventing the curriculum will not make me more interested in learning how my dishwasher works," said social scientist Patti Hausman when speaking to the National Academy of Engineering.[26]

"Feminism as a movement for political and social equity is important, but feminism as an academic clique committed to eccentric doctrines about human nature is not. Eliminating discrimination against women is important, but believing that women and men are born with indistinguishable minds is not," says Pinker.[27] And he is keen to press the case as it fits in well with his thesis that we are not born as blank slates waiting for people to write on. Slipped into the middle of the classic film hit *The Sound of Music* we can see evidence of the extent to which the counter theory that we are shaped by those around us has become part of popular culture. In the opening lines of "Sixteen going on seventeen", the young Austrian soldier Rolfe sings, "You wait little girl on an empty stage; For fate to turn the light on; Your life little girl is an empty page; That men will want to write on." And bleary-eyed Liesl echoes back, "To write on ... "

The answer is that our sexuality is influenced by a little of both – nature and nurture. I don't expect there ever to be a set of data that everyone agrees provides conclusive proof that we are born with an indelibly fixed sexuality, nor that we are totally malleable.

However, the fact that we can't identify something's origins does not falsify its existence. Even though the underlying cause may be difficult to determine, sex differences and individual

---

[26] Quoted in Holden C. (2000), "Parity as a goal sparks bitter battle", *Science*, 289, p. 380.

[27] Pinker, *The Blank Slate*, p. 371.

sexuality are plain to see. I was on holiday in Provence a few years ago and we walked along a hot and dusty pathway to a huge spring – La Fontaine de Vaucluse. In the springtime and autumn some 33,000 gallons of water a second rush up from this swimming-pool-size hole in the ground, but no one knows where the water comes from. Divers have tried to descend to the depths, and explorers have even tried using remote controlled mini submarines equipped with cameras, but still the mystery remains. No one knows the origins of this mighty stream, but equally no one doubts its existence.

Wanting to change

Gender identity is such an inherent and fundamental part of our lives that those whose gender identity coincides with their sexed body cannot possibly understand how anyone might wish to change it. And there lies the crux of the issue – the trans person, whether transexual, transgender or transvestite, is not seeking to change his or her identity at all. What trans people are trying to do is find a way of presenting their gender identity in such a way that the rest of the world will understand who they are.

So says Stephen Whittle in his book *The Transgender Debate*.[28] "It is one thing to say someone has a vulva, vagina, clitoris, breasts, ovaries, etc, it is quite another thing to assume that person is either female, feminine or a woman. Sex is a very complex biological concept which is rarely as unambiguous as we think," he continues.[29]

I want to take a brief look at it because it is yet more proof of the importance that sexuality plays in people's lives.

A famous first occurred in February 1953. When New York photographer George Jorgensen decided to change his name he went a step further. He decided to change his body as well. At

[28] Whittle S. (2000), *The Transgender Debate*, Reading: South Street Press. pp. 2–3.
[29] Whittle, *The Transgender Debate*, pp. 4–5.

the age of 26 the 98-pound ex-GI travelled to Holland for a series of operations that removed his male genitalia and other aspects of his physique such as facial hair. A combination of surgery and hormones then moulded his body, giving it the female appearance that he felt comfortable with. If she had wanted a quiet start to her new life, it was not on offer. Even before she left the Copenhagen hospital as Christine Jorgensen, the news had leaked out, and she was the centre of discussion and debate. "I could never understand why I was receiving so much attention," Jorgensen said in a 1986 interview. "Now, looking back, I realise it was the beginning of the sexual revolution, and I just happened to be one of the trigger mechanisms." Jorgensen lived until he was 62, dying of bladder and lung cancer.

One of the questions frequently asked was whether the operation was a case of "sex reassignment" treatment or an example of "gender confirmation therapy". Were the surgeons giving Jorgensen a new gender, or allowing her to have a body that fitted the gender she had always been? Pioneer endocrinologist Harry Benjamin, who researched into transexualism, said of Jorgensen, "But was this female gender role really new? The vivid description of her early life supplies a negative answer. This was a little girl, not a boy (in spite of the anatomy) who grew up in this remarkably sound and normal family."[30] Since then a few hundred people have "crossed" from one sex to another.

At the moment I don't want to get into a debate about the ethics or politics of the operations, but I raise the issue to point out the extent to which these people feel that their sexuality is a critical part of their identity.

Sexual attraction

There is the all too often corny moment in many a film, and occasionally in real life, where two strangers spy each other on

[30] From the foreword to, Jorgensen C. (1967), *A Personal Autobiography*, New York: Eriksson Inc.

opposite sides of a crowded room, or knock shopping out of each other's arms in a car park, and are inexplicably convinced that they are right for each other. Without even getting around to having a conversation their lives have started to become entwined.

Reality is normally less dramatic, but there is plenty of research looking into what it is that causes particular people to choose each other. As much of it is driven by researchers hell-bent on linking every part of human nature to neo-Darwinistic concepts that competition to pass on our genes is the heart of life, we shouldn't be surprised that much of this work claims to show that attraction between individuals is bound up with the sexual demands of effective reproduction.

There is, for example, evidence that we choose mates who have key differences in the way that their immune system is set up.[31] It seems to work particularly well in mice. Quite how this works in practice is open to conjecture, but the presumed consequence is that any offspring that result from this coupling will have a greater ability to fight disease.

A vast amount of data also shows that we choose mates who are about as good-looking as we are. It's a game you can play at home if you like! Take photographs from a large number of pairs of married people and shuffle them. Place them in two stacks – one of men and the other of the women. Next ask people who don't know them to place the photos in each stack in a row, ensuring that the most good-looking are at one end, the least glamorous are at the other. The result is that when you bring the two rows side by side most of the marriage pairs will be reunited. It appears that when men and women look for partners they make an assessment of their own physical attractiveness and then actively seek a partner with a similar attractiveness "score".[32] It

[31] Stricker L., Amoroso A. and Cerutti N. (1995), "Antigeni de istocompatibilita' epercezione olfattiva nell'uomo", *Antropologia Contemporânea*, 18, pp. 79–86.

[32] Little A. C., Burt D. M., Penton-Voak I. S. and Perrett D. I. (2001), "Self-perceived attractiveness influences human female preferences for sexual dimorphism and symmetry in male faces", *Proc. R. Soc. Lond. B Biol. Sci.* 268(1462), pp. 39–44.

is as if attraction runs by rules of a marketplace. If you have a lot of beauty capital, you can "buy" a high value item.

If beauty is in the eye of the beholder, then the sex of the beholder makes a difference. Men look for aspects of a woman's face that are taken as being typically feminine and vice versa. At root the critical features tend to be the physical aspects that are moulded by the relative levels of the sex hormones testosterone and oestrogen surging through a person's blood vessels. The most prevalent explanation for this is that having strongly male features suggests that you are a man with a solid reproductive capability, and likewise oestrogen-driven characteristics of women's bodies correlate with health and reproductive fitness. Because many researchers see the ability to produce biologically successful offspring as the key influence behind our decisions to form relationships, these evolutionary forces have taught us to take these physical clues seriously. When we come to playing out the aspect of our sexuality that draws us together, it would appear good looks make a difference.[33]

Some researchers claim that we have a sixth sense that is devoted to the task of finding an appropriate mate. While anatomist David L. Berliner was working at the University of Utah in the 1970s he made a chance observation. In the laboratory was a set of jars containing extracts of some tissues. He claims that when he left the jars open, the lab had a puzzling sense of camaraderie, but when he put lids on the jars the friendliness died down. His conclusion was that the tissue samples were releasing pheromones – colourless, odourless chemicals that are known to powerfully influence the behaviour of other animals. These findings led him to look for an organ in humans that could detect these chemicals, and he concluded

---

[33] Perrett D. I., May K. and Yoshikawa S. (1994), "Attractive characteristics of female faces: preference for non-average shape", *Nature*, 364, pp. 239–242; Perrett D. I., Lee K. J., Penton-Voak I., Burt D. M., Rowland D., Yoshikawa S., Burt D. M., Henzi S. P., Castles D. L. and Akamatsu S. (1998), "Effects of sexual dimorphism and facial attractiveness", *Nature*, 394, pp. 884–886.

that the vomeronasal organ (VNO) was responsible. This he describes as a 1cm-long set of ducts that opens into the nasal passage on either side of the central septum. The linings of the ducts appear to contain nerve cells that may be responsible for detecting pheromones and conveying the message to the brain.

That sexually linked chemicals affected hormones was demonstrated by US researcher Martha McClintock who in 1971 showed that female room-mates tended to synchronise their menstrual cycles.[34] Clearly some chemical signal was in operation between them. When I spoke on the phone to Austrian researcher Karl Grammer at the Institute for Urban Ethology in Vienna, he acknowledged that there was a fair degree of scepticism about this sort of work, but insisted that the sceptics are people who don't work with pheromones. "Those who work with them see the evidence and consequently have no doubt that they work," he said. James Kohl, a pheromone researcher on the west coast of America, is equally convinced, saying that there is now good evidence that pheromones can trigger changes in hormone levels. He acknowledges that no one has yet worked out the precise mechanisms, but says that if pheromones work, who cares how they work.

The main point to realise when looking at human sexuality is that it exists and influences our lives. The political and social misuse of difference has made many people wary of discussing issues, and the pornographic use of erotica has made many others feel uncomfortable with even mentioning the word sex. All this is unfortunate, because sexuality is part of who we are as individuals. The diversity of responses to issues that can be categorised as feminine or masculine add to the depth of life. To an extent it matters little whether these differences are

---

[34] McClintock M. K. (1975), "Menstrual synchrony and suppression", *Nature*, 229, pp. 224–225.

biologically or socially induced. The fact that they exist influences who I am as a person, and recognising these differences helps us get a greater insight into what it is to be human and, more specifically, gives us a fuller understanding of what it is to be me.

CHAPTER NINE

a social being

66 "There is no such thing as society." So said the UK's famous Prime Minister, Margaret Thatcher.[1] I am sure that the journalist must have been thrilled by the effect these seven words had, because they were soon echoed around the world by people who thought that this was proof positive that the Iron Lady had gone rusty. It became one of the most talked about phrases of the year. The only shame is that most of these echo generators didn't take time to read the whole of the piece before they extracted the phrase from the page. While it makes a great sound-bite, and cudgel for anyone wanting to beat the political "right", standing alone it does not reflect the real message that she was trying to put over. I have no reason to doubt that Thatcher said the words, nor that the journalist was in any way in error to place them in the piece. But anyone talking into a microphone will say the odd seven words that taken on their own do not represent the total message. It is of secondary importance to note that the interview was for *Woman's Own* magazine, which, while having a reasonably high circulation, is hardly a recognised forum for groundbreaking intellectual debate.

The quote comes from an answer that she gave to a question about what she felt had caused a deterioration in the nation's moral standards.

I think we've been through a period where too many people have been given to understand that if they have a problem, it's the government's job to cope with it. "I have a problem, I'll get a grant." "I'm homeless, the government must house me." They're casting their problem on society. And, you know, there is no such thing as society. There are individual men and women, and there are families. And no government can do anything except through people, and people must look to themselves first. It's our duty to look after ourselves and then, also to look after our neighbour. People have got the entitlements too much in mind,

[1] "Aids, education and the year 2000" (1987), *Woman's Own*, pp. 8–10, 3 October.

without the obligations. There's no such thing as entitlement, unless someone has first met an obligation.

Further on, the article quotes her as saying:

> There is a living tapestry of men and women, and the beauty of that tapestry and the quality of our lives will depend on how much each of us is prepared to take responsibility for ourselves. And each of us, by our own efforts, is prepared to turn round and help those less fortunate.

Taken as a whole, Thatcher's comments in no way deny the existence of society – the beautiful tapestry of living humanity. Instead, they show her understanding that individuals must take responsibility for their own actions if the "living tapestry" of society is going to function well. You can argue the balance of how taxation policy can be adjusted to bring greater peace, depth and beauty to the work of art, and what political and social actions will enable the greatest number of people to lead fulfilled lives, but no one, not even Margaret Thatcher, seriously argues that society does not exist.

The reason for this is that to deny "society" would be to deny one of the critical features of our humanity. We are social beings. The much-quoted phrase of seventeenth-century English poet John Donne (*c.* 1572–1631) neatly, if somewhat morbidly, sums up the idea:

> No man is an island, entire of itself; every man is a piece of the continent, a part of the main. If a clod be washed away by the sea, Europe is the less ... any man's death diminishes me, because I am involved in mankind, and therefore never send to know for whom the bells tolls; it tolls for thee.[2]

A bit of you dies as you stand at a graveside – that is why the vigil hurts.

---

[2] Donne J. (1624), *Devotions upon Emergent Occasions*, "Meditation 17", London. (http://www.ccel.org/d/donne/devotions/devotions-bod.html)

Having said this, the thrust of political policy-making in Europe and America over the last few decades has been to place more and more responsibility on the individual, but this has come at the cost of reducing our expectation of a thriving interaction within society. Having the economic and technical power to supply cars at a price that most families can afford means that we no longer rub shoulders with each other on the way to work. Our desire to live as single entities derives as much from our sense of financial independence as from our determination not to have to be tied down by long-term commitments with others. Wherever possible we seek our independence. Mobile telephones mean that we no longer have to be in any particular place to communicate, which has the knock-on effect that we may talk more often, but meet less frequently. Few people see this as a problem, because the ultimate goal is that "I" do more, and the freedom of movement removes unwanted restraint. Even multichannel television has an effect by removing the corporate aspect of a nation watching a limited repertoire of programmes and commenting on them over the next couple of days; though the mob-following of certain so-called reality-TV programmes where the viewer can become personally involved by submitting their vote shows that people do still enjoy shared experience.

Multiculturalism has also added to the drift away from a social sense of identity. While breaking down barriers that divided groups of people is to be welcomed, this will result in a loss of shared identity unless we create a new concept of our community. The resulting sense of isolation is exacerbated by post-modernity, which claims that notions of definite truth made by twentieth-century scientific modernity can now be questioned. Truth as an absolute concept is now held as being just what you want it to be, which in reality means that there is no truth. This may seem very grown up in that it gives people ownership of their ideas, but it also removes the common ground of shared identity that once was the glue of a community.

In addition, the last century has seen a massive growth of large urban housing estates inhabited by people who have moved into the area, and are not sure how long they will stay. Communication technology lets them stay in touch with old friends, so it feels as if there is less need to make new ones, and rapid transport links to work mean that more and more people commute to work. Their home is more of a dormitory than a centre of community. Consequently, many people share streets with strangers. Changes in working practices also have eradicated the community-building concept of jobs for life. Few people are committed to their company, because they know it could at any moment turn around and kick them out. Instead they are committed to making their own career path as favourable as possible.

This apparently ever-lengthening list can be seen as an inevitable consequence of current lifestyles, but as the growth of the home security industry shows, the twenty-first century is not a place where we feel safe. Our doors have never been stronger nor our windows more secure, but still we install closed circuit television to scrutinise visitors, or try to build gated walls around our little cluster of houses.

Another way of judging whether or not we feel comfortable with this situation is to look at global suicide rates. According to the World Health Organisation the last 45 years have see a 60 per cent increase in worldwide rates of suicide. In 2000, one person committed suicide every 40 seconds, giving an approximate total of one million people for that year alone. Suicide is now the third most common cause of death among those aged between 15 and 44. To gauge the full extent of the way that people are keen to end their lives we should add in the attempted suicides. The WHO says that these occur 20 times more frequently than completed suicides.[3] Interestingly, suicide rates are almost twice as high in Western countries and members of the former soviet block, than in developing countries and South America. Wealth and the trappings of industrial success do not necessarily equate to secure happiness.

[3] http://www5.who.int/mental_health

White South African philosopher Augustine Shutte describes the philosophy underlying the contemporary European concept of a human being.

> Each individual is rather like an atom, separate, autonomous and constrained only by alien forces imposed on it from without. Morality is seen as an essentially private matter ... In this view there is virtually no such thing as a common human nature ... The only thing we have in common is the capacity to originate action, the negative freedom to choose. As such we can of course be the subject of rights, but these rights are not derived from our common human nature. Rather they are produced by agreement of all interested parties.[4]

The "me-centred" idea of community has indeed generated a network of legal "rights". Lawyers and politicians fight over the rulings enshrined in international organisations such as the United Nations Charter on Human Rights, and the European Charter of Human Rights. They have undoubtedly brought a measure of fairness to many situations, but it is important to realise that this is the product of legal negotiation, not the realisation of a fundamental appreciation of what it is to be a human being. When someone says that an action breaches their human right, they may not realise it, but they are making a legal, not an ethical statement.

"Underlying our banking, borrowing, spending and paying of taxes has emerged the unspoken assumption that we function, first and foremost, as *individuals*," say social campaigners Michael Schluter and David Lee in their self-help book on relationships.[5] Their anxiety is that we all too often act as individuals and

---

[4] Shutte A. (1997), "Philosophy for Africa" (paper presented at the University of Cape Town, South Africa), pp. 31–2; quoted in Battle M. (1997), *Reconciliation: The Ubuntu Theology of Desmond Tutu*, Cleveland: The Pilgrim Press.

[5] Schluter M. and Lee D. J. (2003), *The R Option: Building Relationships as a Better Way of Life*, Cambridge: The Relationships Foundation, p. 18.

ignore the fundamental need within each human to live in a network of social connections. The contemporary view, they complain, is that

> "relationship" is a state we may choose to enter into, and choose to withdraw from. Where relationships exist – personal or contractual – we find it increasingly easy to think in terms of what others owe to us rather than what we owe to others. We have rights. We have freedom of choice. We have possessions ... And the mood of this is defensive rather than open ... We teach our children the street-fighting skills of passing examinations, competing for jobs and getting ahead in a career. We don't often pass down to them the relational wealth of valuing and actively cultivating good relationships.

The result is the slow, spiralling weakening of the community we live within – a breakdown in society. I don't mean that we have riots on the streets every night with rape and pillage running as the entertainment of choice for urban thugs. What we have is a feeling that all is not well. You can catch a glimpse of the effect when you note that at a time when living conditions and prosperity have never been so good, the rates of reported depression and suicide are increasing. It is a feeling that runs deep through our very being, and it has its roots in the fact that we are beings that work well with others. Human beings are social animals. Part of what it is to be me is to share my life with others. And this means more than just the tiny number of family members who may share your roof. Sharing your life with others also implies being involved with people who are different from you, share different views from yours, come from different cultures to the one that you feel comfortable in, but are bound together by the realisation that they are human beings.

Yes, if you are wealthy enough, you can build defences around your life to minimise social contact, but this seldom leads to a more fulfilled existence because in doing so you restrict this critical element of your being.

Umuntu ngumuntu ngabantu

The contemporary Western drive for individualism is, however, just one way of looking at life. Another is displayed in the Zulu maxim that states, "*Umuntu ngumuntu ngabantu*", which translates approximately as "You are a person only because of other people".[6] It is a very different form of reductionism to the ones that we are more commonly used to, and it comes from a mindset that is rich in possibility and fires a warning shot across the bow of Western individualism. It draws on the idea that people are defined not by their individual being, by their personal success, wealth or short-term happiness, but by the society they live within – by the people they are a part of. It says that a human being is essentially social.

In the Zulu language the word used to describe humanness is *ubuntu*, and it is represented by equivalent words in the other core African languages.[7] It articulates a world-view, or vision of humanity that has no direct translation into English because it refers to a cultural idea that is far removed from anything we are directly linked to. It can be interpreted as both a factual description and a social ethic to aspire to.

As a part of the underlying thinking, the meaning behind *ubuntu* regards humanity as an integral part of the ecosystem. As such it fits nicely with James Lovelock's Gaia theory (see Chapter 6), but *ubuntu* arrives at this holistic identity not from some principle of science, but from philosophy. "I am an African," writes Thabo Mbeki, President of the African National Congress and President of South Africa, who first shows his relationship to the land before recounting his links to its peoples.

I owe my being to the hills and the valleys, the mountains
and the glades, the rivers, the deserts, the trees, the flowers,

---

[6] Shutte, "Philosophy for Africa".

[7] *Ubuntu* comes from the Nguni language family which comprises Zulu, Xhosa, Swati and Ndebele. *Hunhu* is the equivalent word used in Shona, the other main language of the African continent.

the seas and the everchanging seasons that define the face of our native land. My body has frozen in our frosts and in our latter-day snows. It has thawed in the warmth of our sunshine and melted in the heat of the midday sun. The crack and rumble of the summer thunders, lashed by startling lightning, have been causes both of trembling and of hope ... I am formed of the migrants who left Europe to find a new home on our native land ... I am the grandchild of the warrior men and women ... I am the grandchild who lays fresh flowers on the Boer graves at St Helena and the Bahamas ... I am also able to state this fundamental truth that I am born of a people who are heroes and heroines ... I am born of a people who would not tolerate oppression ... I am born of the peoples of the continent of Africa."[8]

The concept runs deeply through African culture and has become a central cornerstone in thinking in post-apartheid South Africa. At a World Summit held in 2002 in Johannesburg, nations set up huge tented exhibitions displaying all that is good about their country. The South African tent was called the Ubuntu Village. A South African government white paper on welfare states that *ubuntu* is

the principle of caring for each other's well-being ... and a spirit of mutual support ... Each individual's humanity is ideally expressed through his or her relationship with others and theirs in turn through a recognition of the individual's humanity. *Ubuntu* means that people are people through other people. It also acknowledges both the rights and the responsibilities of every citizen in promoting individual and societal well-being.[9]

---

[8] Mbeke T. (1996), A statement on behalf of the African National Congress on the occasion of the adoption by the Constitutional Assembly of the Republic of South Africa Constitution Bill 1996, Cape Town, 8 May; http://www.anc.org.za/ancdocs/history/mbeki/1996/sp960508.html

[9] *South African Government Gazette*, (1996) no.16943, p. 18, paragraph 18, 2 February.

In addition, Ubuntu implies that we have a communal responsibility to sustain life.

A person is a human being because they are enveloped in their community, a community of other human beings. They are caught up in the bundle of life. "To be is to participate," says the internationally famous Archbishop of Cape Town, Desmond Tutu. "The [critical issue] is not independence but sharing, interdependence."[10] What defines a person is their community, not some isolated static quality of rationality, will or memory.

Tutu has his own way of helping people get a feeling of the concept. Here is an example of him telling a parable when he was awarded an honorary degree by Columbia University.

> There was once a light bulb which shone and shone like no light bulb had shone before. It captured all the limelight and began to strut about arrogantly quite unmindful of how it was that it could shine so brilliantly, thinking that it was all due to its own merit and skill. Then one day someone disconnected the famous light bulb from the light socket and placed it on a table and try as hard as it could, the light bulb could bring forth no light and brilliance. It lay there looking so disconsolate and dark and cold – and useless. Yes, it had never known that its light came from the power station and that it had been connected to the dynamo by little wires and flexes that lay hidden and unseen and totally unsung.[11]

"The distinction of each person," says theologian Michael Battle when commenting on Tutu's allegory, "depends upon her or

---

[10] Tutu D. (1973), "Viability" Hans-Jurgen Becten (ed.), *Relevant Theology for Africa*, Durban: Lutheran Publishing House.

[11] Tutu D. (1997), "Response at graduation of Columbia University's honorary doctorate (address to the University of Witwatersrand, 2 August 1982", quoted in Battle, *Reconciliation*, pp. 43–44. The degree was presented by Columbia's president, who had to travel to South Africa with trustees of Columbia University after the South African authorities refused to allow Tutu to travel to the USA.

his connection with other persons and a recognition of a more encompassing context."

Ubuntu underscores the importance of agreement or consensus. An African concept of democracy does not simply boil down to "majority rule" but operates in the form of discussions – and the discussions must be given as much time as is needed to reach a consensus.[12] The agreement, not the time taken, is of paramount importance because it draws from the need to be acting as human being.

Many commentators go on to say that *ubuntu* isn't only concerned with the relationships between people who are currently alive and kicking. For them, the "other people" referred to in the Zulu maxim not only include those who are physically around, but also your ancestors – in African thought, your ancestors are very present members of your extended family. "Not only the living must therefore share with and care for each other, but the living and the dead depend on each other," says South African academic Dirk Louw who works in the Department of Philosophy at the University of the North, 30 km to the east of Pietersburg.[13] At drinking binges known as calabashes, after the word for a local type of beer barrel, revellers pour a little beer on the ground so that their ancestors can join in with the party.

I once saw the belief in the influence of a person's ancestors at first hand when staying with friends in Zimbabwe. They lived on a sugar cane plantation in the southwest of the country, and one day their black Zimbabwean cook disappeared into the undergrowth on all fours. From a Western point of view it looked as if he had had some form of mental breakdown. His friends were anxious for his safety; crawling through snake-filled undergrowth is never a good idea, but they were less perturbed about his state

---

[12] Louw D. J. (2001), "Ubuntu and the challenges of multiculturalism", *Quest*, 15(1–2), p. 19. Published at http://www.eco.rug.nl/cds/Q15.pdf

[13] Louw D. J. "Ubuntu: An African assessment of the religious other", http://www.bu.edu/wcp/Papers/Afri/AfriLouw.htm

of health. "Don't worry," they told us, "he will be fine tomorrow – he has just been visited by his ancestors." The next day he was back at work and nothing was said of the incident.

Respect for other generations has another spin-off. This corporate concept of human identity invokes a concern that human beings need to see natural resources as assets that should be shared on principles of equity, not only among the people who exist at the moment, but also between generations. By this token, abusing a resource for short-term gain is inhumane.

To see how the ideas encompassed by *ubuntu* permeate cultural life within Africa I went and spoke with Paul Bakibinga, a presenter for the African service at the BBC World Service in London. His birthplace was in Uganda where his father had been the principle civil servant in charge of land surveys when British colonialists handed over power in 1962

> I remember my mother talking about *ubuntubulamu*. This is the equivalent word in the Bantu language and expresses the concept of belonging to society, that I am not just an individual but I belong to society – it is something that my mother has been teaching me, or mentioning to me, right from when I was a child.

The individual in a society that is based on the principles described in *ubuntu* is not solitary, but defined in terms of their relationships with others. This means that as relationships change, so too do the individuals. "When my father was alive there would be certain issues I would seek his advice on, such as choosing a wife. Since he has died I would now talk to my brother, he is the heir to my father and has taken that role," Bakibinga told me, adding that his brother would expect to be available to give advice, but that there is no expectation that the brother can force a person to make any particular decision. This changing role has a variety of effects. "Before my father passed away my brother could be my best man at my wedding, but now that he has inherited the role of my father he cannot be my best man because he is in effect my father."

He went on to point out how the concept of interrelationships influences the names given to family members. "Among Busoga,[14] the word for uncle is *babomuto*, 'younger father'," he said. "When my mother refers to my uncles, my father's brothers, she will talk about 'your fathers'. On the maternal side, there is a unification of the sexes. My mother's brother is referred to as my 'mother' – he is taken in the context of the person who gave birth to me. To my sister's children, I am their 'Koja', effectively their younger mother," Bakibinga added with a chuckle.

> My father had two wives. Between him and my mother I have six brothers and one sister, and I have the same numbers of brothers and sisters from my younger-mother. In Africa I would not refer to them as stepbrothers. There is no word for stepbrother. All of my father's children are my brothers and sisters. With my current background and training I sometimes distinguish them, but really there should be no reference at all to that distinction – after all, even the child of a paternal uncle is my brother. In fact there is a case where a judge ruled that there is no such thing as illegitimacy – once you know your father you are no longer illegitimate.

Keeping in touch with all of his "brothers" and "sisters" is an impossible task. Bakibinga's grandfather was chief of the area and so he had 27 wives. "Even today I meet new brothers [or cousins in European speak] that I have never met before," he laughed.

The use of these words does more than indicate which part of the family everyone belongs to; they also show lines of authority. Your father's brother is your "younger father" because he has a shared authority over you. It is his job to ensure that a youngster is brought up in a disciplined manner, and he will feel a father's pride when the child does well.

The sense of shared responsibility within a family has other implications. A father's eldest sister is called a *senga*. It is the *senga*'s

---

[14] Busoga are a Bantu ethnic group whose language is Lusoga.

role, among other things, to teach young daughters about sex and how to behave at home. As is often the case, the importance of any concept is often revealed most starkly when it breaks down or starts to fail. Africa has seen an incredible mass movement of people from tribal communities to cities. Bakibinga's parents were born and brought up in Kamali, a village 80 miles from Uganda's capital city Kampala, but he was born and raised in the city. This social transition means that a youngster's *senga* is often no longer around. Consequently there is no one to educate girls, because discussing sex with your parents is still counted as taboo.

Bakibinga believes that this is one of the factors that has led to a rise in promiscuity, which in turn has poured oil on the fire of HIV-AIDS that is currently running through Africa. The problem is exacerbated in that HIV-AIDS has also killed many *sengas*, taking away the very people who could be in the best position to give life-saving advice. "In fighting AIDS we have been trying to find a new person to play the role of the *senga*, the role of the aunt in society," explained Bakibinga, who has worked on various anti-AIDS initiatives over the last few years. He says that some social groups are trying to encourage young girls to take advice from someone who, although not immediately related to the family, can play the role of *senga*. Part of the thinking underlying this drive is to reawaken the notion of *ubuntu*, in that your life as part of the human community stretches beyond the family. Accepting this paves the way for girl's *senga* to come from outside her immediate family.

One illustration of this wider implication of *ubuntu* can be seen in one of the incidents Bakabinga told me about. At one point he was working on an assignment in Uganda and needed to take a long taxi ride. As they passed through one village some children were playing dangerously close to the roadside.

The driver took it upon himself to stop the car and hit the kids – not in a vicious way – but it was part of his responsibility to discipline the kids and to say "keep away from the road because it is dangerous." And when this particular driver had administered the corporal punishment I heard the mother saying "yes, that is good – they have told you – I have always told you it is bad to play by the road." That was a sense of belonging to a community.

Bakibinga contrasts this integrated view of a caring society with the life he experiences in London.

In London I don't know who my neighbour is. I come in. I lock my door. If anything happens to me my neighbour is not the first person I refer to or call on if I miss any groceries or run out of salt. Even in an African city you could always run to the neighbour and they would be happy to help you. I can't get that here because society is very much individualistic.

But let's not get too carried away. If taken as the only measure by which we define people and their role within society, *ubuntu* too has problems. The most obvious being that it can enforce uniformity on to a group. "In short," says Louw,

although it articulates such important values as respect, human dignity and compassion, the *ubuntu* desire for consensus also has a potential dark side in terms of which it demands an oppressive conformity and loyalty to the group. Failure to conform will be met by harsh punitive measures.[15]

To apply *ubuntu* this rigidly would deny that human beings are also individuals, and as such are essentially free to come to their own opinions, and bear responsibility for their actions and outcome. But as a component of the way we see ourselves, as a facet of our humanity ubuntu has much to teach us.

A political panacea

There appears, however, to be a paradox. If *ubuntu* is so powerful at binding people together in harmony, why is violence

---

[15] Louw, "Ubuntu and the Challenges", p. 19.

apparently endemic in Africa? For example, in the South African province of KwaZulu-Natal, where *ubuntu* is claimed to be part of every day life, violent ethnic and political clashes still occur frequently and this is far from the only example of conflict on the continent of Africa.

"The apparent anomaly posed by the occurrence of such violent conflicts, significantly fades once one concentrates on the many counter examples. African examples of caring and sharing, and of forgiving and reconciliation abound. Ask any South African," challenges Louw, explaining that the transition from an oppressive apartheid regime in South Africa to one with a firmer understanding of democracy for all human beings living in that country has been vastly more peaceful than anyone could have predicted.[16] Many commentators have attributed this to the underlying ethos of *ubuntu*, and in particular to the way that people like Tutu helped to apply it at a political level. *"Ubuntu* is especially appealed to when it comes to the settlement of seemingly unsolvable conflicts and insurmountable contradictions."[17]

The key to the peaceful revolution was a realisation that being linked together, no one benefits from violent retribution. In a book charting the application of *ubuntu* to this era of history, theologian Michael Battle says that Tutu built a theological model of *ubuntu*. In this he sought to show that if you released an oppressed person, you would restore the oppressor's humanity. By liberating South African black people, you would give the powerful whites a renewed awareness of their humanity. He encouraged everyone to see the oppressed and their oppressors as equal beings under God. "For Tutu, *ubuntu* expresses this mutuality. The relationship of oppressor and oppressed and the resulting definition of humanity through racial classification are

[16] Louw, "Ubuntu: An African assessment".

[17] Binsbergen W van. (2001) "Ubuntu and the globalisation of Southern African thought and society", *Quest*, 15(1–2), p. 74. Published at http://www.eco.rug.nl/cds/Q15.pdf

broken through *ubuntu*, an alternative way of being in a hostile world," comments Battle.[18]

As he drew his Nobel Prize lecture to a close on 11 December 1984, peace laureate Tutu said, "Perhaps oppression dehumanizes the oppressor as much as, if not more than, the oppressed. They need each other to become truly free, to become human. We can be human only in fellowship, in community, in *koinonia*,[19] in peace."[20] For Tutu, this interrelationship is not just a way of resolving conflict – it is a way of life, because a self-sufficient human being is subhuman.

> I have gifts that you do not have, so, consequently, I am unique – you have gifts that I do not have, so you are unique. God has made us so that we will need each other. We are made for a delicate network of interdependence. We see it on a macro level. Not even the most powerful nations in the world can be self-sufficient.[21]

"Ubuntu," says Battle,

> refers to the person who is welcoming, who is hospitable, who is warm and generous, who is affirming of others, who does not feel threatened that others are able and good for [this person], has a proper self-assurance that comes from knowing they belong in a greater whole, and knowing that they are diminished when another is humiliated, is

---

[18] Battle, *Reconciliation*, p. 5.

[19] *Koinonia* is a Greek word used in the biblical New Testament and is translated: "To have fellowship ... to have a share in, to participate in, and to be in communication with, someone or something", *New Dictionary of Christian Ethics and Pastoral Theology*, Leicester: Intervarsity Press, 1995, p. 379. As such it can be used for the fellowship of people operating as an effective team in the workplace.

[20] Tutu D. Nobel lecture. http://www.nobel.se/peace/laureates/1984/tutu-lecture.html

[21] Tutu D. (1997), "God's dream", in Kreiger D. and Kelly F. (eds.), *Waging Peace II: Vision and Hope for the 21st Century*, Chicago: The Nobel Press, n.d.; quoted in Battle, *Reconciliation*, p. 35.

diminished, is tortured, is oppressed, is treated as if they were less than who they are. What a wonderful world it can be, it will be, when we know that our destinies are locked inextricably into one another's.[22]

What an *ubuntu* understanding of humanity brought to South Africa was the possibility of forgiveness rather than retribution. It provided a rationale for realising that while grave wrongs had been committed, these would not be solved or laid to rest by violence. Such application of force, while potentially understandable, would hurt individual human beings, and as such would harm everyone. It would only serve as a form of self-mutilation. Instead of lengthy lawsuits that would have been unlikely to result in any form of genuine justice, South Africa instigated its truth and reconciliation tribunals, where people could effectively confess to their part in the previous regime and in doing so enable a painful line to be drawn under their past.

"Many Western views of personhood," says Battle,

> center on the lone individual, whose essential characteristic is that of self-determination, whereas the African view of a person depicts the person's meaning or intelligibility only in the context of his or her environment. In the African concept of *ubuntu*, which Tutu appropriated for his own purposes, human community is vital for the individual's acquisition of personhood.[23]

And this idea has had radical results. "The relatively non-violent transition of the South African society from a totalitarian state to a multi-party democracy, is not merely the result of the compromising negotiations of politicians," says Louw. "It is also – perhaps primarily – the result of the emergence of an ethos of

[22] Battle, *Reconciliation*, p. 35.

[23] Battle, *Reconciliation*, p. 37.

solidarity, a commitment to peaceful co-existence amongst ordinary South Africans in spite of their differences."[24]

Ubuntu serves as a cohesive moral value in the face of adversity, argues president of the Philosophical Society of Southern Africa Joe Teffo. Although the policy of apartheid greatly damaged the overwhelming majority of black South Africans,

> there is no lust for vengeance, no apocalyptic retribution ... A yearning for justice, yes, and for release from poverty and oppression, but no dream of themselves becoming the persecutors, of turning the tables of apartheid on white South Africans ... The ethos of *ubuntu* ... is one single gift that African philosophy can bequeath on other philosophies of the world.[25]

Another commentator, Siphisa Maphisa, puts it this way:

> South Africans are slowly re-discovering their common humanity. Gone are the days when people were stripped of their dignity (*ubuntu*) through harsh laws. Gone are the days when people had to use *ubulwane* [i.e. animal-like behaviour] to uphold or reinforce those laws. I suggest that the transformation of an apartheid South Africa into a democracy is a re-discovery of *ubuntu*.[26]

Life's tapestry is full of colour and the pattern created as we mingle together varies from place to place around the world, as well as altering though history. There is, however, no time that

[24] Louw, "Ubuntu: An African assessment".

[25] Teffo, J. (1994), *The Concept of Ubuntu as a Cohesive Moral Value*, Pretoria: Ubuntu School of Philosophy, p. 5. Quoted in http//www.bu.edu/wcp/Papers/Afri/AfriLouw.htm

[26] Maphisa S. (n.d.), *Man in Constant Search of Ubuntu: A Dramatist's Obsession*, Pretoria: Ubuntu School of Philosophy. p. 8. Quoted in http//www.bu.edu/wcp/Papers/Afri/AfriLouw.htm

archaeologists and anthropologists can point to at which human beings are existing as solitary animals, fending for themselves in lonely isolation. You can try all you like to be independent, but as soon as you turn on a light bulb, draw water from a tap or walk to the shops you become part of the greater mob of human social interaction. You may not want to take things to the extreme suggested in *Ubuntu* theory, but there is no getting away from the fact that we live within societies, and that as such the sum of the whole is greater than that of the constituent parts. We are, after all, social beings.

free to be me

I sat in my car outside a school recently – I was driving through a strange town and stuck in a traffic jam. It was home time, and young children and teenagers were pouring out through the main gate and heading home in the summer sun. Some kept themselves to themselves with their hands thrust in pockets and their heads bowed; trying to avoid any gaze. Others jostled along in clusters, walking in step and seemingly all talking at once. Most were dressed in the regulation uniform dictated by school rules: grey trousers or skirt, white shirt, grey-and-red stripy tie, black shoes and a dark red blazer with a golden yellow logo woven into the breast pocket.

Some looked keen, alert and ready for the evening ahead. Others slumped along, seemingly shattered by whatever this hot and humid day had dished out to them. A crowd gathered some distance down the road waiting for a bus, while others ran to cars and drove off with parents. A group of parents stood at one side, many pushing child buggies, ready to walk home with the younger children.

I scrutinised the faces of the people streaming past. I knew none of them, not as people, and would probably never meet any of them again, and if I did I would not be aware of it. To me, they were impersonal beings. I half-closed my eyes and scanned the scene, reducing this flow of people to a simple mass of humanity. It was much like looking at an ants' nest and seeing the overall bustle of activity, but not focusing on any individual ant. But then between the eyelashes I saw the odd blurs of colour. They turned out to be the sports bags hung over some of the youngsters' shoulders and presumably stuffed with a mixture of textbooks and sweaty sports kit. These seemed to be attempts at establishing some form of individuality while not stepping too far out of line and breaching the well-known dress-code conformity. The occasional pupil took this a step further and sported flashes of bright colours dyed into their hair, or heavy chained jewellery that was hastily being placed back in position now that they had escaped from the strict confines of the school grounds and could reassert what they felt was the part of their identity that they wanted to portray.

We live our lives as one of the mass, but are simultaneously aware that each of us is unique. We want to be "me", and to be treated as "me" by society; to be given space and permission to chase personal dreams, while at the same time wanting to be free from any feeling of threat, coercion or intimidation from others.

I want to be free – free to be me. But here we face the problem that surfaces in many a teenage mind and is often suppressed, but seldom goes away: What is me? It is at that point that we start using different categories to aid some form of definition, either for ourselves or others. I'm a powerful sportsman; she's just like her mother; he's a Muslim extremist ... the list is endless.

Through this book we have taken a look at nine different aspects of life that can be used to build a picture of individuals. I could have generated more, but had to stop somewhere. The nine chosen are an introductory set and not intended to be seen as an inclusive, all-encompassing list. While the categories have their use to a point, in that they can encourage us to investigate one particular aspect of a person's make-up, they become dangerously restrictive if the analysis ends there. The sportsman is also material, spiritual and sexual; the young woman is not only genetically influenced by her mother, but is material, related and historic. The Muslim extremist is more than spiritual, he is social, conscious and physically embodied.

Each of the people interviewed for this book may be discussed in one chapter, but that is purely an artificial division. They could easily appear in other sections. Arthur White was certainly physical, but also strongly aware of the spiritual part of his nature as well as the importance to him of his history in London's East End. His consciousness collects objective information and makes frequent subjective assessments. He is aware of his sexuality, seeing how this influences his attractiveness and vanity. He knows that his genes gave him a head start in the strength game, but that his personal desire to win has played an important role in his life.

In itself, taking a holistic view of our existence is not particularly radical. But it is radically different from the way we have started to think and behave. All too often we are encouraged to think of ourselves in terms of single categories. This short-sighted view of humanity is very flat, and right now, science is being used to iron out even more of the interest. Certain voices spend a great deal of time telling us that science can define us without reference to anything else. We are told that science shows that we are not as free to be me as we would like to believe, but instead that we are regulated and predictable. We are materially and genetically preordained to be the way we are. All you can do is live with it.

All too readily we move from the scientific desire to measure, test, define and categorise, to a mindset that says we are determined by these categories. And there are two branches of science in particular that are currently making that sort of claim – genetics and neuroscience.

In March 2003, Nobel Prize-winner Paul Nurse suggested to an audience at the Royal Society that we were only a few years away from each child being given a "genetic identity card" at birth.[1] He warned that this could lead to genetic apartheid as insurers and employers rule out people with particular genetic "defects". This, however, will be the case only if we allow ourselves to buy the narrow-minded view that we are our genes. Yes, genetic screening will give much valuable information, but it will be a huge mistake to overemphasise its use because it can only look at the genetic part of our make-up and leaves the rest unassessed. The genetic ID will be problematic if employers start to believe it is the final say – it will be restrictive if insurance companies begin to believe that genes rule our lives.

It shouldn't really need saying, but science itself shows us that this genetic reductionism is massively over-simplistic. David Barker's work, for example, demonstrates that genes are only one aspect of our inherited make-up, and that any assessment of how those genes are put to work to create each individual person

---

[1] Reported by the BBC, "Warning on gene 'ID cards'". 4 March 2003; http://news.bbc.co.uk/2/hi/health/2816003.stm

needs a careful consideration of their history. More recently an experiment that produced cloned cats discovered that the kittens ended up looking very different from each other. Clearly their genes were only part of the story.[2]

In terms of neuroscience, the 2003 BBC Reith Lectures gave an interesting example of how people try to produce single-item definitions of human beings, but how easily these fall to pieces. In the first of the series of lectures entitled *The Emerging Brain*, neuroscientist Vilayanur Ramachandran set out his stall with the following statement;

> Even though it is common knowledge these days, it never ceases to amaze me that all the richness of our mental life – all our feeling, our emotions, our thoughts, our ambitions, our love life, our religious sentiments and even what each of us regards as his own intimate private self – is simply the activity of these little specks of jelly in your head, in your brain. There is nothing else.[3]

The initial statement went down well within the cosy confines of the reductionally converted who populated the auditorium, where people are prepared to turn a blind eye to these weaknesses. But it would have a much more limited resonance in the real world, where people realise that we are more than can be made sense of by any study of the material nature of our brain.

But even Ramachandran couldn't keep the argument of "nothing else" for long. Two weeks later, when talking about people's appreciation of art he said:

> Let us assume that the variance you see in art is driven by cultural diversity or – more cynically – by just the auctioneer's hammer, and only 10% by universal laws that

---

[2] Highfield, R. (2003), "First cloned cat definitely not the spit of her mother", *Daily Telegraph*, 23 January.

[3] Ramachandran V. S. (2003), "Lecture 1: phantoms in the brain", *BBC Reith Lectures*. www.bbc.co.uk/radio4/reith2003

are common to all brains. The culturally driven 90% is what most people already study – it's called art history. As a scientist what I am interested in is the 10% that is universal – not in the endless variations imposed by cultures.[4]

Even by his own appraisal of the situation, which you could argue would be biased to give the brain the biggest role possible, when it came to art, he had to admit that neuroscience would only explain one-tenth of the story.

It is also important to be honest about the limited scope of science. Plenty is written about the undoubted wonders of science, and rightly so, because without science we would have little of the technology that makes our lives richly fulfilling and safe. But it would be a mistake to think that science has the last word in human capability, and is the ultimate intellectual technique that will reveal all. It would be an error to expect science to explain me.

The error lies at two levels. The first mistake is to believe that science arrives at definite conclusions. In reality, good experiments tend to raise more questions than answers. Yes, our area of knowledge is increasing, but not as fast as the area of things that we know we don't know – the zone of our known ignorance. As Austrian-born, British philosopher and statistician Karl Popper put it, "Our knowledge can only be finite, while our ignorance must necessarily be infinite."[5] Science and scientists will never be able to know everything, and most of what we call knowledge would more truthfully be described as "best working model". While this may well generate technologies that are useful, it is a long way from suggesting that the underlying science is infallible.

In a different territory of science I noticed that physicist Stephen Hawking has moved away from a belief that he and

---

[4] Ramachandran V. S. (2003), "Lecture 3: The artful brain", *BBC Reith Lectures.* www.bbc.co.uk/radio4/reith2003

[5] Popper K. (1963), *Conjectures and Refutations*, Routledge.

his colleagues would come up with a single theory capable of explaining everything. At the end of his bestseller, *A Brief History of Time*, he had written of his expectation of finding the unified theory, "for then we would know the mind of God". Now he says, "Maybe it is not possible to formulate the theory of the Universe in a finite number of statements."[6] Hawking now suggests that science may not be capable of producing the answers it is looking for.

The second error is to believe that science has got further than it has. If, for example, you have been following the story of the human genome project, you may be surprised to know that the full sequence of genetic code in the centre of our cells has not been fully sequenced. Yes, in June 2000, Prime Minister Blair and President Clinton held simultaneous press conferences announcing the successful competition of this milestone in human endeavour, but what they were really talking about was the first draft of most of the sequence. In February 2001 there was another fanfare. Now it really was finished – well, most of it – at least enough of it to make a reasonable job of starting to analyse the sequence. But if you didn't listen very carefully, you would have come away with the idea that the sequence was complete – job done. Then in April 2003 we had yet another announcement. The "human genome project is finished – really". Speaking at the party in April 2003 to mark the event, Francis Collins, director of the National Human Genome Research Institute in the USA, likened the achievement of sequencing the entire human genome to scaling Mount Everest. But yet again, this was not all that it seemed. All the bits of the genetic sequence that were easy to get at were done, but a remaining 1 or so per cent that was "intractable to sequencing" was left undone. It strikes me that this would have been similar to a mountain climber coming down from Everest saying that he had completed the mountain, only to reveal on questioning that

<hr>

[6] Quoted in Brooks M. (2003), "The impossible puzzle", *New Scientist*, p. 34, 5 April.

he had done all except the top 1 per cent, the really difficult bit, which was intractable to assent.

But we are all too ready to put reality aside. There is still a tendency to hold the general opinion that a scientific explanation has the last word, and in the context of this book, many voices continue to claim that science can produce watertight descriptions of what it is to be human.

More than me

Once we start to take seriously a multifaceted view of human beings it has implications throughout all sorts of areas of policy. Areas of life and death are obvious examples. As we saw in the Introduction to the book, a pattern of thought that says "You are your consciousness" ignores any other aspect of your existence. It may be the case that a patient in a coma has massively disabled consciousness but they are still part of a network of family relationships, have a history, are a material and embodied being. Any decision made purely on the basis of the patient's consciousness, that ignores all other elements, will be in danger of ignoring many other aspects of that person. It might make a doctor's job easier to focus on one measurable aspect of a person's existence and base pronouncements on that, but easier is not necessarily better.

Over the last couple of years we have seen the extent to which the embodied part of our existence is viewed increasingly as being important in the outcries against the way that some hospitals have treated dead bodies. For example, there was outrage after a hospital in Liverpool revealed that its staff had retained organs from dead babies without telling parents,[7] and a feeling of revulsion after bacon was found placed on the body of a Muslim woman in a hospital morgue.[8] Without thinking about it, we know that we are more than just our conscious, self-aware brain.

[7] For example "Organ scandal families accept £5m", 31 January 2003; http://news.bbc.co.uk/1/hi/england/2713479.stm

[8] "Dead body defiled in hospital", 17 April 2003; http://news.bbc.co.uk/health

Likewise, a holistic approach to euthanasia probably makes the issue more complex, but more honest. The simplistic view is that killing yourself is OK, because you are simply exercising your right of autonomy. Why should anyone stop you if you want to end your life? A holistic view reveals that you are also part of your family. Your death will harm others. It will arguably also harm society because we are social beings and the death of one person leaves an unfillable space. You could even argue that in asking for suicide, a person takes away another's opportunity to serve – service may seem gruelling, but it is at the heart of community.

The whole area of disability then comes into view. We commonly look at people as if they are defined by their physical abilities. Traditionally we haven't expected any wheelchair-user to get a place at a university or to want to go shopping. Consequently most of our public spaces are inaccessible to them or, at best, highly inconvenient. But people with disabilities are every bit spiritual, conscious and social. If we remembered to focus on these elements of people's beings when we design facilities there would be a greater chance of naturally creating environments that are inclusive of people with all levels of ability.

A holistic view of humanity would also show that many more of us are disabled than we might initially think. If we are concerned only about our children's physical fitness, but ignore their spiritual growth then we are liable to produce children with a spiritual disability. If we concentrate on helping people with their sexuality, but ignore their ability to develop social skills we will find ourselves in a society that is relationally disabled. We should see the protection of our material world, the environment, as being just as important as looking after our appearance or keeping in touch with relations.

You could also apply the holistic approach to questions about designer babies. Yes, using technology to manipulate someone's genes will have a profound influence over them, but no amount of genetic control is going to dictate the results. Why not? Because that is only one aspect of our existence. Try as they might, any control freak parent is unlikely to be able to exert the scale of influence required to produce a musical child or an artistic or mathematical genius. There is too much we don't understand, and too many aspects to regulate at once.

The division between humanity and all other animals also starts to make more sense if you take a holistic approach. This way there is little surprise that animals share most, if not all, of our features and abilities, but generally do so to incredibly limited extents. It's interesting to watch a Disney cartoon and see what happens to animals in order to bring them to life as "people" – suddenly they express conscious subjective opinions, they plan ahead with purpose, have enhanced abilities to act socially and can perform physical tasks that are impossible given the structures of their bodies.

Most religions give explanations for the difference between human beings and the rest of the animal kingdom that run along the lines of "God says so" or "God made it that way". But outside this, the holistic approach points a way forwards. No individual aspect is absolutely unique to humans, but the way we employ each ability is vastly more complex, so much so that when added together there is clear blue water between us and any other species. You could argue that it is slightly messy, because it suggests that there is no killer fact that ends all arguments, but it also implies that we have no carte-blanche ability to use other species at will, as bad use is abuse. When all things are added together, we are more than just animals.

What it is to be human

A limitation of this book is that the choice of aspects chosen is my own personal subjective choice. You could easily argue that some of the chapters are less important than others, and that some aspects of what it is to be human are missing. It would have been easy to include a chapter on language, showing that the uniquely capable way that we communicate with each other is more than a physical ability, but also influences the way we live and to an extent changes the way that we view the world. I could have considered memory, looking maybe at people who have lost their memory as a way of studying how our recall of personal experience affects our present.

This limitation, however, this is also a strength. No matter how many chapters were included in the book, there would always be more. Some may overlap with, or be a subdivision of, others, but all the same you could put forward a legitimate argument for their inclusion. It might well be that if I set out to write the book now, I would choose a slightly different set of chapters to illustrate the range of aspects that go into making us who we are.

At the same time, the people chosen to appear in each chapter are there precisely because they are extreme examples of each aspect of life. This choice is deliberate. The task of each chapter has not been to argue that each part is vitally important to everyone all the time, but to look at what a person's life looks like when that aspect becomes dominant. Joanna Rose and Christine Whipp feel that the lack of information about their ancestral background has radically influenced their lives. As such they act as examples of people who feel that their family background is part of who people are. For other people born in similar circumstances, however, the situation would be less extreme, so much so that some would even argue that it was irrelevant. In reality I do not believe that any of the aspects covered are irrelevant for anyone, but that individual characteristics may currently play a reduced role in their lives.

In the Introduction we took a prism and broke up a beam of white light into its constituent parts. We have taken a similar approach to analysing ourselves. We can split up being human into categories that are to an extent artificial, but at the same time they are useful. There is no clear boundary between a person's embodied being and their sexuality; there is no demarcation between their genetic make-up and their physical body; between their consciousness and their social behaviour. But using the categories makes us aware of our complexity – it forces us away from reductionist simplicity, and the fun starts when we recombine the categories, when we reintroduce the colours.

I enjoy going to the theatre. I find it much more alive than cinema, and much more uncertain. There is more of an element of risk. Sometimes you come out shocked into silence by the brilliance of the performance, and on other occasions you can't wait to get home. On these occasions I find myself studying the scenery and the lighting. I used to love building stage sets and enjoy seeing how people with massive resources go about the task. The lighting is fascinating because of the games you can play with it. Blast a stage with white light and it will look very flat, the actors will look ill. Cover it with a mixture of different colours and you introduce warmth and a feeling of depth, though the overall effect is that the stage looks bathed in white. Music hall stages in fact used this principle to the full. They simply washed the stage with red, blue and green lights. Turn them all on together and you have white; dim some and brighten others and you can mix any colour you fancy to change the mood of a scene.

So too the human condition. People vary in mood and personal colour when different aspects of their lives shine to different degrees of intensity. By this analogy, you could surmise that a fully balanced person who has all aspects of life running perfectly and at full power would be like a white light, and I doubt such a person exists. The real world is occupied by people who have some parts of their life running well and other aspects have faded almost to non-existence. The differences add colour to life. At the same time there is no reason to suggest that a person's "colour" might not change throughout life as different parts of their make-up become more or less important. It is a process that we should look for and encourage.

Any approach to define who I am is doomed from the outset because we will always be more complex than can be catered for by any single definition. It is, however, not intractable to investigation once you have broken the subject down into bite-sized pieces, and have admitted that revealing the nature of one aspect of our existence is not the same as describing humanity. This antidote to reductionism lets us glimpse the wonder of being me.

Bibliography

Battle M. (1997), *Reconciliation: The Ubuntu Theology of Desmond Tutu*, Cleveland: The Pilgrim Press.

Bauby J-D. (1997), *The Diving Bell and the Butterfly*, Fourth Estate.

Brevard A. (2000), *The Woman I Was Not Born to Be: A Transsexual Journey*, Philadelphia: Temple University Press.

Colapinto J. (2000), *As Nature Made Him: The Boy Who Was Raised a Girl*, Quartet Books.

Dawkins R. (1976), *The Selfish Gene*, Oxford University Press.

Djerassi C. (2001), *This Man's Pill: Reflections on the Fiftieth Birthday of the Pill*, Oxford University Press.

Du Boulay S. (1988), *Tutu: Voice of the Voiceless*, Penguin.

Du Boulay S. (2001), *Facing Disfigurement with Confidence*, A Changing Faces Publication.

Fukuyama F. (2002), *Our Posthuman Future: Consequence of the Biotechnology Revolution*, Profile Books.

Gray J. (1993), *Men Are from Mars, Women Are from Venus*, HarperCollins.

Habgood J. (1998), *Being a Person*, Hodder & Stoughton Religious.

Hallowell E. M. and Ratey J. J. (1994), *Driven to Distraction: Recognizing and Coping with Attention Deficit Disorder from Childhood through Adulthood*, New York: Simon & Schuster.

Hay D. with Nye R. (1998), *The Spirit of the Child*, London: Fount.

Jones S. (1993), *The Language of the Genes*, Flamingo.

Lovelock J. (2000), *The Ages of Gaia: A Biography of Our Living Earth*, Oxford University Press.

Lovelock J. (2000), *Homage to Gaia: The Life of an Independent Scientist*, Oxford University Press.

Lovelock J. (2000), *A New Look at Life on Earth*, Oxford University Press.

Malik K. (1996), *The Meaning of Race: Race, History and Culture in Western Society*, Basingstoke: Palgrave.

Malik K. (2000), *Man, Beast and Zombie: What Science Can and Cannot Tell Us about Human Nature*, London: Phoenix.

McCloskey D. N. (1999), *Crossing: A Memoir*, The University of Chicago Press.

McFadyen A. I. (1990), *The Call to Personhood: A Christian Theory of the Individual in Social Relationships*, Cambridge University Press.

Muggeridge M. (1981), *Ancient and Modern*, London: British Broadcasting Corporation.

Pinker S. (2002), *The Blank Slate: The Modern Denial of Human Nature*, Allen Lane.

Rose S., Lewontin R. C. and Kamin L. J. (1984), *Not in Our Genes: Biology, Ideology and Human Nature*, Penguin.

Schluter M. and Lee D. (1993), *The R Factor*, London: Hodder & Stoughton.

Schluter M. and Lee D. (2003), *The R Option: Building Relationships as a Better Way of Life*, Cambridge: The Relationship Foundation.

Shakespeare T. (2000), *Help*, Venture Press.

Singer P. (1994), *Rethinking Life and Death*, Oxford University Press.

Stock G. (2002), *Redesigning Humans: Choosing Our Children's Genes*, Profile Books.

Vardy P. and Grosch P. (1994), *The Puzzle of Ethics*, London: HarperCollins-Religious.

Warnock M. (1998), *An Intelligent Person's Guide to Ethics*, Gerald Duckworth & Co.

Warnock M. (2001), *What Is It to Be Human? What Science Can and Cannot Tell Us*, Institute of Ideas.

White A., Johnson S. and McDowall I. with Murray M. (2000), *Tough Talk: True Stories of East London Hard Men, E-bouncers and Debt Collectors*, Word Publishing.

Whittle S. (2000), *The Transgender Debate*, Reading: South Street Press.

Wright L. (1997), *Twins: Genes, Environment and the Mystery of Identity*, Phoenix.

Zemen A. (2002), *Consciousness: A User's Guide*, Yale University Press.

Index

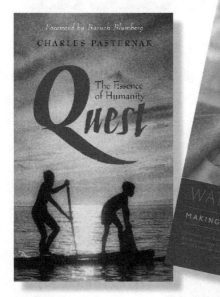